PRACTICAL STATISTICS FOR NURSING AND HEALTH CARE

PRACTICAL STATISTICS FOR NURSING AND HEALTH CARE

Jim Fowler

Principal Lecturer,
Department of Biological Sciences, De Montfort University, Leicester, UK

Phil Jarvis

Consultant Statistician,
Enabling Science & Technology, Astra Zeneca Pharmaceuticals, Macclesfield, UK

and

Mel Chevannes

Professor of Nursing and Head of Department of Nursing,
De Montfort University, Leicester, UK.

JOHN WILEY & SONS, LTD

Copyright © 2002 by John Wiley & Sons Ltd,
Baffins Lane, Chichester,
West Sussex PO19 1UD, England

National 01243 779777
International (+44) 1243 779777
e-mail (for orders and customer service enquiries): cs-books@wiley.co.uk
Visit our Home Page on http://www.wiley.co.uk or http://www.wiley.com

Other Wiley Editorial Offices

John Wiley & Sons, Inc., 605 Third Avenue,
New York, NY 10158–0012, USA

Wiley-VCH Verlag GmbH, Pappelallee 3,
D-69469 Weinheim, Germany

John Wiley & Sons (Australia) Ltd, 33 Park Road, Milton,
Queensland 4064, Australia

John Wiley & Sons (Asia) Pte Ltd, 2 Clementi Loop #02–01,
Jin Xing Distripark, Singapore 0512

John Wiley & Sons (Canada) Ltd, 22 Worcester Road,
Rexdale, Ontario M9W 1L1, Canada

British Library Cataloguing in Publication Data
A catalogue record for this book is available from the British Library

ISBN 0 471–49715 0 (Hardback)   0–471 49716 9 (Paperback)

Typeset by Kolam Information Services Pvt. Ltd, Pondicherry, India
Printed and bound in Great Britain by TJ International Ltd, Padstow, Cornwall
This book is printed on acid-free paper responsibly manufactured from sustainable forestry, in which at least two trees are planted for each one used for paper production.

CONTENTS

PREFACE   xi

FOREWORD TO STUDENTS   xv

1 INTRODUCTION   1

1.1 What do we mean by statistics?   1
1.2 Why is statistics necessary?   1
1.3 The limitations of statistics   2
1.4 Calculators and computers in statistics   2
1.5 The purpose of this text   3

2 HEALTH CARE INVESTIGATIONS: MEASUREMENT AND SAMPLING CONCEPTS   5

2.1 Introduction   5
2.2 Populations, samples and observations   5
2.3 Counting things – the sampling unit   6
2.4 Sampling strategy   7
2.5 Target and study populations   8
2.6 Sample designs   8
2.7 Simple random sampling   9
2.8 Systematic sampling   9
2.9 Stratified sampling   10
2.10 Quota sampling   11
2.11 Cluster sampling   12
2.12 Sampling designs – summary   12
2.13 Statistics and parameters   13
2.14 Descriptive and inferential statistics   13
2.15 Parametric and non-parametric statistics   14

3 PROCESSING DATA   15

3.1 Scales of measurement   15
3.2 The nominal scale   15
3.3 The ordinal scale   16
3.4 The interval scale   17

3.5   The ratio scale   17
3.6   Conversion of interval observations to an ordinal scale   17
3.7   Derived variables   19
3.8   Logarithms   20
3.9   The precision of observations   21
3.10 How precise should we be?   22
3.11 The frequency table   22
3.12 Aggregating frequency classes   24
3.13 Frequency distribution of count observations   26
3.14 Bivariate data   27

4   PRESENTING DATA   29

4.1   Introduction   29
4.2   Dot plot or line plot   29
4.3   Bar graph   30
4.4   Histogram   32
4.5   Frequency polygon and frequency curve   33
4.6   Scattergram   35
4.7   Circle or pie graph   35

5   CLINICAL TRIALS   39

5.1   Introduction   39
5.2   The nature of clinical trials   39
5.3   Clinical trial designs   40
5.4   Psychological effects and blind trials   41
5.5   Historical controls   42
5.6   Ethical issues   43
5.7   Case study: Leicestershire Electroconvulsive Therapy (ECT) study   43
5.8   Summary   45

6   INTRODUCTION TO EPIDEMIOLOGY   47

6.1   Introduction   47
6.2   Measuring disease   48
6.3   Study designs – cohort studies   50
6.4   Study designs – case-control studies   51
6.5   Cohort or case-control study?   53
6.6   Choice of comparison group   54
6.7   Confounding   55
6.8   Summary   56

7   MEASURING THE AVERAGE   57

7.1   What is an average?   57
7.2   The mean   57

7.3	Calculating the mean of grouped data	59
7.4	The median – a resistant statistic	60
7.5	The median of a frequency distribution	61
7.6	The mode	62
7.7	Relationship between mean, median and mode	64

8 MEASURING VARIABILITY 65

8.1	Variability	65
8.2	The range	65
8.3	The standard deviation	66
8.4	Calculating the standard deviation	67
8.5	Calculating the standard deviation from grouped data	68
8.6	Variance	69
8.7	An alternative formula for calculating the variance and standard deviation	70
8.8	Obtaining the standard deviation and sum of squares from a calculator	71
8.9	Degrees of freedom	71
8.10	The Coefficient of Variation (CV)	72

9 PROBABILITY AND THE NORMAL CURVE 75

9.1	The meaning of probability	75
9.2	Compound probabilities	76
9.3	Critical probability	78
9.4	Probability distribution	79
9.5	The normal curve	81
9.6	Some properties of the normal curve	82
9.7	Standardizing the normal curve	83
9.8	Two-tailed or one-tailed?	84
9.9	Small samples: the t-distribution	86
9.10	Are our data 'normal'?	88
9.11	Dealing with 'non-normal' data	91

10 HOW GOOD ARE OUR ESTIMATES? 95

10.1	Sampling error	95
10.2	The distribution of a sample mean	95
10.3	The confidence interval of a mean of a large sample	98
10.4	The confidence interval of a mean of a small sample	99
10.5	The difference between the means of two large samples	100
10.6	The difference between the means of two small samples	102
10.7	Estimating a proportion	103
10.8	The finite population correction	105

11 THE BASIS OF STATISTICAL TESTING   107

11.1 Introduction   107
11.2 The experimental hypothesis   107
11.3 The statistical hypothesis   108
11.4 Test statistics   110
11.5 One-tailed and two-tailed tests   110
11.6 Hypothesis testing and the normal curve   111
11.7 Type 1 and type 2 errors   113
11.8 Parametric and non-parametric statistics: some further
      observations   113
11.9 The power of a test   114

12 ANALYSING FREQUENCIES   115

12.1 The chi-squared test   115
12.2 Calculating the test statistic   115
12.3 A practical example of a test for homogeneous frequencies   118
12.4 One degree of freedom – Yates' correction   119
12.5 Goodness of fit tests   120
12.6 The contingency table – tests for association   121
12.7 The 'rows by columns' ($r \times c$) contingency table   125
12.8 Larger contingency tables   127
12.9 Advice on analysing frequencies   129

13 MEASURING CORRELATIONS   131

13.1 The meaning of correlation   131
13.2 Investigating correlation   131
13.3 The strength and significance of a correlation   133
13.4 The Product Moment Correlation Coefficient   134
13.5 The coefficient of determination r^2 136
13.6 The Spearman Rank Correlation Coefficient r_s 137
13.7 Advice on measuring correlations   139

14 REGRESSION ANALYSIS   141

14.1 Introduction   141
14.2 Gradients and triangles   142
14.3 Dependent and independent variables   143
14.4 A perfect rectilinear relationship   144
14.5 The line of least squares   146
14.6 Simple linear regression   147
14.7 Fitting the regression line to the scattergram   150
14.8 Regression for estimation   150
14.9 The coefficient of determination in regression   151
14.10 Dealing with curved relationships   152
14.11 How we can 'straighten up' curved relationships?   155
14.12 Advice on using regression analysis   155

15   COMPARING AVERAGES   157

15.1	Introduction	157
15.2	Matched and unmatched observations	158
15.3	The Mann–Whitney U-test for unmatched samples	158
15.4	Advice on using the Mann–Whitney U-test	160
15.5	More than two samples – the Kruskal–Wallace test	161
15.6	Advice on using the Kruskal–Wallace test	163
15.7	The Wilcoxon test for matched pairs	164
15.8	Advice on using the Wilcoxon test for matched pairs	167
15.9	Comparing means – parametric tests	168
15.10	The z-test for comparing the means of two large samples	168
15.11	The t-test for comparing the means of two small samples	170
15.12	The t-test for matched pairs	171
15.13	Advice on comparing means	173

16   ANALYSIS OF VARIANCE – ANOVA   175

16.1	Why do we need ANOVA?	175
16.2	How ANOVA works	176
16.3	Procedure for computing ANOVA	178
16.4	The Tukey test	181
16.5	Further applications of ANOVA	183
16.6	Advice on using ANOVA	185

APPENDICES   187

Appendix 1: Table of random numbers	187
Appendix 2: t-distribution	188
Appendix 3: χ^2-distribution	189
Appendix 4: Critical values of Spearman's Rank Correlation Coefficient	190
Appendix 5: Critical values of the product moment correlation coefficient	191
Appendix 6: Mann–Whitney U-test values (two-tailed test)	192
Appendix 7: Critical values of T in the Wilcoxon test for matched pairs	193
Appendix 8: F-distribution	194
Appendix 9: Tukey test	198
Appendix 10: Symbols	200
Appendix 11: Leicestershire ECT study data	201
Appendix 12: How large should our samples be?	203

BIBLIOGRAPHY	209
INDEX	211

PREFACE

Background

Over the past few years, substantial changes have taken place in the education, training and practice of nursing and midwifery. All pre-registration education and training is now provided at a minimum of Diploma of Higher Education, and much is at degree level. Pre-registration education and training incorporates an equality of hours between knowledge (2300 hours) and clinical/practice skills (2300 hours). Knowledge includes the application and use of scientific skills (ENB 1999; DoH 1993; 1999)*. Additionally, there are many continuing professional development and post-graduate training opportunities designed to enable practitioners to maintain their knowledge and skills.

More recently, the UKCC (1999) emphasized the effects of health service changes on the work of nurses and midwives:

> '... future of the health services raised a number of paradoxes likely to affect the role of nurses and midwives; in particular the greater demands upon nurses and midwives for technical competence and scientific rationality'.

Pre-registration education and training and continuing professional development are required to prepare nurses and midwives with technical and scientific skills to work competently in the NHS today and in the future. These skills demand an understanding of the principles underlying statistics, and using them appropriately in the daily practice of nursing and midwifery.

*Details of references cited throughout this book are listed in the bibliography in Appendix 13.

Changes in the context of health care

The context of health care is changing, most notably between primary care/home/community-based services and those provided in acute hospitals. Public expectation of wellness and patterns of disease have also changed (DoH 1999). At the same time, opportunities created by the development in pharmaceutical services and technological advances combine to amplify the dynamic context of health care where nurses, midwives and other health professionals work. All these changes point to a need for nurses and midwives to be prepared to respond in knowledgeable and practical ways, and to do so competently every time.

Technological imperatives

Telemedicine and telecare will make screening for diseases, diagnosis and monitoring quicker and more accurate. Patients and other users are likely to benefit from an improvement in the methods by which diagnoses are made, and as a result, care and treatment will be more effective (DoH 1999). Nurses and midwives need to know and understand the data they see, the most appropriate analysis to make, and how to use the information in the patient's care and treatment plan.

Team working

Organizational changes in the way nurses work make it imperative to do so in a team and to provide an integrated planning pathway of care for patients. A Plan of Care, for example for patients with diabetes, may include evaluating response to medication – insulin or tablets – by age, gender and ethnicity, or undertaking a comparable analysis of the stabilization of diabetes among a group of patients by insulin injection.

Nurses working in specialist areas such as heart, thoracic surgery, intensive care, renal, endosocopy clinics, and cancer centres/units use skills of a highly technical and statistical nature. Students and qualified nurses in these settings should be enabled to access and manage health information, including the accurate recording of clinical changes in individual patients' conditions (ENB 1999). Analysis of clinical variations by drawing on basic statistical principles enable health care professionals to decide on appropriate treatment and predict recovery.

In relation to midwifery, the Institute of Manpower Studies (1993) found that over 40 percent of maternity services had recently introduced 'team midwifery'. The practice of team midwifery requires all midwives to use scientific evidence to under-pin the care they provide to women and babies. For example, an understanding of the AGPAR score of the baby at birth and in the immediate post-delivery period is important to a team of midwives operating in the delivery room, and the post-natal area. A baby's APGAR score at birth, and in particular when there is delay, with the maximum score of 10 being achieved postnatally may indicate the need for accurate monitoring of heart, pulse and breathing rates.

Modern ways of working to make a difference

The new education and training of nurses and midwives spelt out in *Making a Difference* (DoH 1999) emphasizes modern roles in acute, community and primary care. Nurses, midwives and health visitors working in the community and in primary care are expected to provide accurate and reliable health information to patients about health risks and patterns of disease. As public health workers, health visitors under-take community profiling, health needs assessment, and provide advice about a range of topics, including the control of communicable diseases.

Requirement of health visitors to prescribe drugs (from the Nurses' Prescribing Formulary for patients) places additional demands for good statistical understanding and skills in order to prescribe accurately, appropriately and safely. As this role expands in the future, and for nurses to administer a range of medicines with direct referral to a doctor (DoH 2001), the need for the use of quantitative methods in nursing will increase.

In the context of care for young children, health visitors and paediatric nurses know that mothers are likely to be anxious about whether the weight of their child is high, low or just within the expected range. The ability of these health professionals to understand the principles underlying weight calculations, analysis and predictions place them in a good position to accurately explain and reassure mothers.

This book on practical statistics for nurse and other health care professionals is an introductory text produced to assist with the growing demand for user-friendly worked examples which are clear and easily understood, not only by students, but also by busy practitioners. We aim to provide an introduction to the statistical techniques that are most commonly used that also provides a sound foundation for those who choose to expand their careers into research.

Jim Fowler
Phil Jarvis
Mel Chevannes

FOREWORD TO STUDENTS

It is possible that you have recently enrolled on a nursing, midwifery or other health care course. It is likely that you did so with high ideals and a motivation to help humankind in the relief of suffering. It is also possible that you did not enjoy handling 'numbers' at school, and you have sought to escape from them. And now you have in your hands a textbook that might, at first sight, look terrifying.

We wish to offer a word of reassurance. Modern health care is a 'science', and as such, involvement with numbers is inescapable. It was, after all, Florence Nightingale herself who recognized the importance of maintaining accurate numerical records. However, it might not be as bad as you fear. This book seeks to guide you through the subject 'from scratch', and we make no assumptions about your previous learning. The chapters develop in difficulty progressively through the book, and the material is extensively cross-referenced. If you wish to 'dip in' to the text, you will find guidance on where to look back for under-pinning explanations.

It is very unlikely that you will ever need to know all that is contained in this book. In the first instance, it may be simply a supporting text to help you through a statistical element of your course. Later, you may be involved in a project of some kind, when this book can help you plan correctly the gathering, presentation and analysis of your data. Some of you may then venture into an area of research, in which case the more advanced chapters in this book will give you a sound foundation in the quantitative techniques that are required.

We plead with you not to feel intimidated by the formulae that you see by flicking through these pages. How they are used is carefully described in each case. By persisting with the book from 'square one', reworking some of our own examples to make sure that you get the same answer, you will rapidly become sufficiently confident to apply them to your own data. And, who knows, you may even come to enjoy statistics!

1 INTRODUCTION

1.1 What do we mean by statistics?

Statistics are a familiar and accepted part of the modern world, and already intrude into the life of every nurse and health care worker. We have statistics in the form of patients registered at a GP practice or outpatient clinic; hospital measurements and records of temperature, blood pressure and pulse rate; data collected from various surveys, censuses and clinical trials, to name but a few. It is impossible to imagine life without some form of statistical information being readily at hand.

The word **statistics** is used in two senses. It refers to collections of quantitative information, and methods of handling that sort of data. A hospital's database listing the names, addresses and medical history records of all its registered patients, is an example of the first sense in which the word is used. Statistics also refers to the drawing of inferences about large groups on the basis of observations made on smaller ones. Estimating the relationship between smoking and the incidence of lung cancer illustrates the second sense in which the word is used.

Statistics looks at ways of organizing, summarizing and describing quantifiable data, and methods of drawing inferences and generalizing upon them.

1.2 Why is statistics necessary?

There are two reasons why some knowledge of statistics is an important part of the competence of every health care worker. First, statistical literacy is necessary if they are to read and evaluate reports and other literature critically and intelligently. Statements like 'there was a significant improvement in mean scores on the depression scale over time, $F_{(2,92)} = 13.99$, $P < 0.001$' (Cimprich & Ronis 2001), enable the reader to decide the justification of the claims made by the particular author.

A second reason why statistical literacy is important to health care workers is if they are going to undertake an investigation that involves the collection, processing and analysis of data on their own account. If the results are to be presented in a form that will be authoritative, then a grasp of statistical principles and methods is essential. Indeed, a programme of work should be planned anticipating the statistical methods that are appropriate to the eventual analysis of the data. Attaching some statistical treatment as an afterthought to make a survey seem more 'respectable' is unlikely to be convincing.

1.3   The limitations of statistics

Statistics can help an investigator describe data, design experiments, and test hunches about relationships among things or events of interest. Statistics is a tool that helps acceptance or rejection of the hunches within recognized degrees of confidence. They help to answer questions like, 'If my assertion is challenged, can I offer a reasonable defence?'

It should be noted that statistics never *prove* anything. Rather, they indicate the likelihood of the results of an investigation being the product of chance.

1.4   Calculators and computers in statistics

A hand calculator that has the capacity to calculate a mean and standard deviation (typically referred to as a 'scientific' calculator) from a single input of a set of data is indispensable. All the calculations and worked examples in this book were first worked out using such a calculator. Because different makes of calculator operate somewhat differently, we have not attempted to offer guidance about the use of individual calculators: we suggest that you study the instruction booklet that comes with your calculator.

In a modern world, computer packages are readily available and easy to use. However, we suggest caution against jumping straight into computer packages without first understanding the underlying background and principles of a particular statistical technique. Computers undertake any analysis that you ask of it, but can not provide the intelligent reasoning about whether the test is appropriate for the kind of data you are using. Moreover, a 'print-out' of the analysis can be ambiguous and confusing if you do not understand the underlying principles. We feel this is best achieved by first familiarizing yourself with the techniques 'long hand', working through our own examples and applying them to your own data. In due course, support from a computer package will become a natural extension of your analysis.

1.5   The purpose of this text

The objectives of this text stem from the points made in Section 1.2. First, the text aims to provide nurses and health care workers with sufficient grounding in statistical principles and methods to enable them to read survey reports, journals and other literature. Secondly, the text aims to present them with a variety of the most appropriate statistical tests for their problems. Thirdly, guidance is offered on ways of presenting the statistical analyses, once completed.

Full details of references and other material that we suggest for further reading are listed in full in the Bibliography in Appendix 13. For assistance in cross-referencing, we classify items according to chapter. Thus, Section 9.1, Figure 9.1, Table 9.1 and Example 9.1 are all to be found in Chapter 9.

2 HEALTH CARE INVESTIGATIONS: MEASUREMENT AND SAMPLING CONCEPTS

2.1 Introduction

A health care investigation is typically a five-stage process: identifying objectives; planning; data collection; analysis; and, finally, reporting. The methodologies frequently used are sample surveys, clinical trials and epidemiological studies that are the subject of this and subsequent chapters. However, we must first be clear about the definitions of some basic terms. Many of the terms used in statistics also have usage in daily life, where the meaning might be quite different. The word 'population' may conjure images of 'people', whilst 'sample' might mean a 'free sample' of cream offered by a pharmaceutical company, or a 'sample' requested by a doctor for urine analysis. In statistics, however, these words have much more precise meanings.

2.2 Populations, samples and observations

In statistics, the term 'population' is extended to mean any collection of individual items or **units** that are the subject of investigation. Characteristics of a population that differ from individual to individual are called **variables**. Length, age, weight, temperature, number of heart beats, to name but a few, are examples of variables to which numbers or values can be assigned. Once numbers or values have been assigned to the variables, they can be measured.

Because it is rarely practicable to obtain measures of a particular variable from all the units in a population, the investigator has to collect information from a smaller group or **sub-set** that represents the group as a whole. This sub-set is called a **sample**. Each unit in the sample provides a record, such as a measurement, which is called an **observation**. The relationship between the terms we have introduced is summarized below:

Observation:	3.62 kg
Variable:	weight
Sample unit (item):	a new-born male baby
Sample:	those new-born male babies that are weighed
Statistical population:	all new-born male babies that are available for weighing.

Note that the *biological* or *demographic* population would include babies of both sexes, and indeed, all individuals of whatever age or sex in a particular community.

2.3   Counting things – the sampling unit

We sometimes wish to count the number of items or objects in a group or collection. If the number is to be meaningful, the dimensions of the collection have to be specified.

For example 'the number of patients admitted to an accident and emergency department' has little meaning unless we know the time scale over which the count was made. A collection with specified dimensions is called a **sampling unit**. An observation is, of course, the number of objects or items counted in a sampling unit. Thus, if 52 patients are admitted to a particular A & E department in a 24 hr period, the sampling unit is 'one A. & E. 24-hour period' and the observation is 52. The sample is the number of such 24-hour periods that were included in the survey. However, the definition of the 'population' requires care. It might be tempting to think that the population under investigation is something to do with patients, but this is not the case when they are being counted. The statistical population comprises the same 'thing' as the sample units that comprise the sample. In this case, the statistical population is a rather abstract concept, and represents all possible 'A & E department 24 hour periods' that *could* have been included in the survey.

It is very important to be able to identify correctly the population under investigation, because this is essential in formulating a 'null hypothesis' when undertaking statistical tests. This is the subject of Chapter 11.

2.4 Sampling strategy

As we said above, it is not always possible or practicable to sample every single individual or unit in a particular population either due to its size, or constraints on available resources (for example, cost, time, manpower). The solution is to take a sample from the population of interest and use the sample information to make inferences about the population.

A common, but misguided, approach to sampling is to first decide what data to collect, then undertake the survey, and finally, decide what analyses should be done. However, without initial thought being given to the aims of the survey, the information or data may not be appropriate (e.g. wrong data collected, or data collected on wrong subjects, or insufficient data collected). As a result, the desired analysis may not be possible or effective.

The key to good sampling is to:

1. Formulate the aims of the study.

2. Decide what analysis is required to satisfy these aims.

3. Decide what data are required to facilitate the analysis.

4. Collect the data required by the survey.

The crucial point relates to the sequence. For example, if the aim of a study is to identify the effectiveness of asthmatic care within a single GP practice, suitable measures of effectiveness need to be defined. One measure could be based on the number of acute asthma exacerbations (deteriorations) in the preceding 12 months, and this number could be compared with that for the previous 12 months. Other measures might assess the number of patients who have had their inhaler technique checked or are using peak flow meters at home. Most of this information can be obtained from practice records, although cross-checking with hospital records may be required to validate the assessment based on acute exacerbations.

2.5   Target and study populations

We have to distinguish between the target and study populations. The **target population** in the asthma example above is the number of patients registered with the GP practice who have asthma. The **study population** consists of all patients who could actually be selected to form the sample, i.e. those who are *known* to have asthma. For example, a proportion of the target population may not know they have asthma, will not therefore be registered, and thus will not form part of the study population. Ideally, the 'target' and 'study' populations coincide.

2.6   Sample designs

Once the study population has been defined, the next task is to decide which subjects from the population should form the sample. The following list is not exhaustive, but gives a selection of sample designs pertinent to audit:

- simple random sampling
- systematic sampling
- stratified sampling
- quota sampling
- cluster sampling.

The first three designs can be applied to sampling from finite populations, i.e. situations where every member of the study population can be identified. Such is the case in our asthmatic care example (Section 2.4), where a list of all asthmatic patients registered with the GP practice is available or can easily be obtained prior to the study. Quota and cluster sampling are used when it is not possible or practicable to enumerate every member of the study population.

2.7 Simple random sampling

In a **simple random sampling** design, every individual in the study population has an equal chance of being included in the sample. That is to say, steps are taken to avoid *bias* in the sampling. In our asthma example above, the population being sampled is all patients registered with the GP practice who are known to have asthma (say, 800). To select a simple random sample of size $n = 20$, each patient ('sampling unit') is assigned a unique number: 1, 2, 3, and so on, until all 800 patients have been numbered. Then 20 numbers in the range 1 to 800 are selected at random, and the patients (sampling units) corresponding to these numbers represent the sample.

There are two usual ways of obtaining random numbers. First, many calculators and pocket computers have a facility for generating random numbers. These are often in the form of a fraction, e.g. 0.2771459. You may use this to provide a set of integers, 2, 7, 7, 1, ...; or 27, 71, 45, ...; or 277, 145; or 2.7, 7.1; and so on, according to your needs, keying in a new number when more digits are required.

Secondly, use may be made of **random number tables**. Appendix 1 is such a table. The numbers are arranged in groups of five in rows and columns, but this arrangement is arbitrary. Starting at the top left corner, you may read: 2, 3, 1, 5, 7, 5, 4...; or 23, 15, 75, 48, ...; or 231, 575, 485...; or 23.1, 57.5, 48.5, 90.1, ...; and so on, according to your needs. When you have obtained the numbers you need for your investigation, mark the place in pencil. Next time, carry on where you left off. It is possible that a random number will prescribe a subject (sampling unit) that has already been drawn. In this event, ignore the number and take the next random number. The purpose is to eliminate *your* prejudice as to which items should be selected for measurement. Unfortunately, **observer bias,** conscious or unconscious, is notoriously difficult to avoid when gathering data in support of a particular hunch!

Random sampling is the preferred approach to sampling. Although it does not *guarantee* that a representative sample is taken from the study population (due to *sampling error*, described in Section 10.1), it gives a better chance than any other method of achieving this.

2.8 Systematic sampling

Systematic sampling has similarities with simple random sampling, in that the first subject in the sample is chosen at random and then every subsequent tenth or twentieth patient (for example) is chosen to cover the entire range of the population.

Example 2.1

What interval is required to select a systematic sample of size 20 from a population of 800?
The required fixed interval is:

$$\frac{\text{population size}}{\text{intended sample size}} = \frac{800}{20} = 40$$

Therefore, the first patient ('sampling unit') is selected at random (as described in Section 2.8) from among patients numbered 1 to 40. Suppose number 23 is selected. The sample then comprises patients 23, 63, 103, 143, ..., 783.

A disadvantage of systematic sampling occurs when the patients are listed in the population in some sort of periodic order, and thus we might inadvertently systematically exclude a subgroup of the population. For example, given a population of 800 patients listed by 'first attendance' at the clinic, and that over a 20 week period, 40 patients registered per week, 20 during the daytime and 20 during the evening surgeries. If these patients were listed in the following order: Week 1 daytime patients, Week 1 evening patients, Week 2 daytime patients, ..., Week 10 evening patients, then selecting patients 23, 63, ..., 783 would result in a sample of evening clinic patients, and exclude all the daytime patients. It is possible that this could generate a biased, or unrepresentative, sample.

An argument in favour of systematic sampling occurs when patients are listed in the population in chronological order, say, by date of first attendance at the GP practice. A systematic sample would yield units whose age distribution is more likely to perfectly represent the study population.

2.9   Stratified sampling

Stratified sampling is effective when the population comprises a number of subgroups (or 'sub-populations') that are thought to have an effect on the data being collected, such as male and female, different age groupings, or ethnic origin. These subgroups are called **strata**. A **stratum** ('layer') is defined as a collection of individuals (sampling units) that are as alike as possible. For example, the credibility of results from a study of breast cancer would be in doubt if the proportion of pre-menopausal patients differed between two samples selected for comparison. By defining two strata, namely 'pre-menopausal patients' and 'not pre-menopausal patients', this problem is avoided.

A simple random sample is taken from each stratum. The resulting *stratified samples* are then more likely to reproduce the characteristics of the population. The two main approaches to deciding how many individuals should be sampled from each stratum are *equal allocation* and *proportional allocation*. The first approach results in an equal number of individuals per stratum, while the second provides samples in which the sample sizes from each stratum reflects the sizes of those in the population.

2.10   Quota sampling

Quota sampling differs from stratified sampling in that a simple random sample is not chosen from each stratum. Instead, the sample is obtained by using the most accessible patients, as long as they represent the identified subgroups. For example, if we require details relating to 20 women patients with asthma between 30 and 50 years of age, we do not identify all individuals satisfying these criteria in the population in order

to take a simple random sample of these. Rather, we simply select the first 20 individuals who present themselves and fulfil these criteria.

Quota sampling is so called because the number of sampling units (e.g. patients) required in a particular sample is referred to as the **quota** to be obtained. If making comparisons between different sub-groups (e.g. adults and children), the sizes of the sample from each subgroup are usually decided to reflect the proportions in the population. For example, if there are twice as many adults as children in the available population, the quota of adults is twice as large as the children.

The main problem with quota sampling is that accessible individuals may not be representative of the study population. Patients who attend at their GP practice regularly may be different to those who don't, or who are unable to attend through work or other commitments.

2.11   Cluster sampling

Cluster sampling involves dividing the population into subgroups called clusters. However, unlike stratified sampling and quota sampling (in which the subjects in a particular stratum or subgroup are meant to be as alike as possible), the objective is to include in each cluster the various characteristics that the population might contain. The rationale for both stratified and quota sampling is the control of factors (e.g. age or sex differences) that are known (or suspected) to confound the response being investigated. In cluster sampling, the idea is not to have a homogeneous group, but one which is representative of the cluster through either a census (100% sample) or, more usually, by taking a representative sample of the cluster.

Cluster sampling is commonly used when the population covers an area that can be divided by region (e.g. GP practices). A small number of these clusters is selected at random (using simple random sampling). Every subject in the chosen clusters is then included in the sample. One key problem with cluster sampling is choosing appropriate clusters.

2.12   Sampling designs – summary

Choice of correct survey method is extremely important. The best approaches to sampling from a finite population, as in our asthma example, are to use either a simple random sample or a stratified random sample. Stratification is used when it is known that the response of interest is related to some factor (e.g. age or sex).

Choice of appropriate sampling method is not always obvious, and may involve a mixture of the methods we have described. Always seek advice if you are in doubt, as the cost of advice in relation to the cost of obtaining the sample is very small. Scheaffer *et al.* (1990) provide a very good introduction to all aspects of survey sampling.

.13   Statistics and parameters

Measures that describe a variable of a sample are called **statistics**. It is from the sample statistics that the **parameters** of a population are estimated. Thus, the average weight of a sample of new-born male babies is the statistic which is used to estimate the average weight (parameter) of a population of new-born male babies. An easy way to remember this is: *Statistics is to sample as parameter is to population.*

Sometimes populations appear to be rather abstract or hypothetical concepts, in which case their parameters are also hypothetical. We can calculate an 'average temperature' from a sample of ten observations collected from a patient over a day. What exactly is the parameter that this statistic is estimating? It is the hypothetical 'population' of all temperature observations that *could* be made during the observation period.

In estimating a population parameter from a sample statistic, the number of observations in a sample can be critical. Some statistical methods depend upon a minimum number of sampling units, and where this is the case, it should be borne in mind before commencing your study. Whilst it is true that larger samples will invariably result in greater statistical confidence, there is nevertheless a 'diminishing returns' effect. In many cases the time, effort and expense involved in collecting very large samples might be better spent in extending the study in other directions. We offer guidance as to what constitutes a suitable sample size for each statistical technique as it arises.

.14   Descriptive and inferential statistics

Descriptive statistics are used to organize, summarize and describe measures of a sample. No predictions or inferences are made regarding population parameters. **Inferential** (or **deductive**) **statistics**, on the other hand, are used to infer or predict population parameters from sample measures. This is done by a process of inductive reasoning based on the mathematical theory of **probability**. Fortunately, only a very minimal knowledge of mathematical theory of probability is needed in order to apply the rules of the statistical methods, and the little that is needed will be explained. However, no-one can predict exactly a population parameter from a sample statistic, but only indicate with a stated degree of confidence within what range it lies. The degree of confidence depends upon the sample selection procedures and the statistical techniques used.

2.15   Parametric and non-parametric statistics

Statistical methods commonly fall into one of two classes – **parametric** and **non-parametric**. Parametric methods are the oldest, and although most often used by statisticians, may not always be the most appropriate for analysing medical data. Parametric methods make strict assumptions that may not always hold true.

More recently, non-parametric methods have been devised which are not based upon stringent assumptions. These are frequently more suitable for processing the sort of data we collect. Moreover, they are generally simpler to apply since they avoid the laborious and repetitive calculations involved in some of the parametric methods, although, as we note elsewhere, computers that are used intelligently can help with this. The circumstances under which a particular method should be used will be described as it arises. A summary showing which methods should be applied in particular circumstances is provided in Section 11.8.

3 PROCESSING DATA

3.1 Scales of measurement

Variables measured by nursing and other health care workers can be either **discontinuous** or **continuous**. Values of discontinuous variables assume integral whole numbers, and are usually **counts** of things (frequencies). On the other hand, values of continuous variables may, in principle, fall at any point along an uninterrupted scale, and are usually measurements (length, weight, temperature, pH, etc.). Measurement values may sometimes appear to be integral whole numbers if the recorder elects to measure to the nearest whole unit; this does not, however, obviate the fact that there can be intermediate values. The distinction between 'count data' and 'measurement data' is an important one that we shall often refer to.

Generally, four levels of measurement are recognized. They are referred to as **nominal, ordinal, interval** and **ratio** scales. Each level has its own rules and restrictions; moreover, each level is hierarchical in that it incorporates the properties of the scale below it.

3.2 The nominal scale

The most elementary scale of measurement is one that does no more than identify categories into which individuals or items may be classified. The categories have to be mutually exclusive, in other words it should not be possible to place an individual into more than one category. For example, gender, age-class, ethnic origin, type of disease (e.g. category of cancer) and blood groups A, B, AB and O are all nominal categories into which count data can be uniquely assigned.

The *name* of a category can of course be substituted by a *number*, but it will be a mere label and have no numerical meaning. Thus, non-smokers, cigarette smokers, cigar smokers and pipe smokers are nominal categories into which counts of people can be assigned. We can label these categories, respectively, 1, 2, 3 and 4. But the numbers are just codes or labels, and have no mathematical significance.

3.3   The ordinal scale

The ordinal scale incorporates the classifying and labelling function of the nominal scale, but in addition brings to it a sense of order. Ordinal numbers are used to indicate **rank order**, but nothing more. The ordinal scale is used to arrange (or rank) individuals into a sequence ranging from the highest to the lowest, according to the variable being measured. Ordinal numbers assigned to such a sequence may not indicate absolute quantities, nor can it be assumed that intervals between adjacent numbers are equal. Categories of patient age might be the basis of an ordinal scale:

	Score
Child	1
Teen	2
Adult	3
Older adult	4
Elderly	5
Older elderly	6
Frail elderly	7

In this scale there is no simple relationship between the numerical values on the age scale. 'Older elderly' does not mean twice 'Adult', but it will always be ranked above 'Elderly'.

Another well-known ordinal scale is the **AGPAR Scale**, developed in 1952 as a measure of the physical condition of a new-born baby. An AGPAR score is derived from summing five observations relating to heart rate, respiratory effort, muscle tone, reflex irritability and colour. Each component is rated on a three-point scale, 0, 1 and 2. Therefore, the final AGPAR score lies between a minimum of 0 (all components score 0) and a maximum of 10 (all components score 2). The assessment is conducted 1 minute after birth, and four minutes later. Clearly, a baby with a score of 0 five minutes after birth has, by implication, no measurable heart-beat or respiration, and would thus be considered to be in critical condition.

3.4 The interval scale

As the term **interval** implies, in addition to rank ordering data, the interval scale allows the recognition of *precisely how far apart* are the units on the scale. Interval scales permit certain mathematical procedures untenable at the nominal and ordinal levels of measurement. Because it can be concluded that the difference between the values of, say, the 8th and 9th points on the scale is the same as that between the 2nd and 3rd, it follows that the intervals can be added and subtracted. But because a characteristic of interval scales is that they have *no absolute zero point* it is *not* possible to say that the 9th value is three times that of the third. A common interval scale is temperature: 10°C is not twice as 'hot' as 5°C because the zero on the scale in question (Celsius) is not absolute (the absolute zero on that scale is of course −273°C).

3.5 The ratio scale

The highest level of measurement, which incorporates the properties of the interval, ordinal and nominal scales, is the **ratio** scale. A ratio scale includes an absolute zero, it gives a rank ordering and it can simply be applied for labelling purposes. Because there is an absolute zero, all of the mathematical procedures of addition, subtraction, multiplication and division are possible. Measurements of length and weight fall on ratio scales. Thus, a length of 150 mm *is* three times as long as one of 50 mm.

The mathematical properties of interval and ratio scales are very similar, and because no statistical procedure described in this book distinguishes between them, we shall refer to them both simply as 'interval' scales.

3.6 Conversion of interval observations to an ordinal scale

Usually, observations made on interval scales allow the execution of more sensitive statistical analysis. Sometimes, however, interval data are not suitable for certain methods. Perhaps because the data are too skewed, we are forced to downgrade them to an ordinal rank scale for use in non-parametric methods. The following measurements (mm) are ranked in increasing size in the top line. Their rank (ordinal) scores are underneath:

Length (mm): 31.0 31.4 32.3 33.1 33.5 34.9 35.0 36.6 37.2 38.0

Rank score: 1   2   3   4   5   6   7   8   9   10

If large numbers of observations are collected, it is inevitable that some of the observations will be equal in value. Their ranks will also be tied, and these have to be dealt with correctly. Since some statistical tests which we describe later depend upon the ranking of observations, we take the opportunity now of dealing with the problem of **tied observations**.

Where tied observations occur, each of them is assigned the average ranks that would have been assigned if there had been no ties. To illustrate this, a set of measurements rounded to the nearest whole number is given below. For convenience they are presented in ascending order and adjacent tied scores are underlined:

25 26 27 27 28 29 30 30 30 31 32 33 33 33 33
34 35 36 36 36 36 36 37

If we try to rank these, the single extreme values of 25 and 37 will clearly be ranked 1 and 23, respectively. The two values of 27 together occupy the ranks of 3 and 4; they are each assigned the average rank of $3\frac{1}{2}$. The three ranks of 30 occupy the ranks of 7, 8 and 9. They have an average rank of 8. In similar manner, the four values of 33 are each assigned the rank $13\frac{1}{2}$, and the five of 36 the rank of 20. The set of data is rewritten below, with the correct ranks assigned:

Observation: 25 26 27   27 28 29 30 30 30 31 32

Rank: 1   2   3½ 3½   5   6   8   8   8 10 11

Observation: 33   33   33   33   34 35 36 36 36 36 36 37

Rank: 13½ 13½ 13½ 13½ 16 17 20 20 20 20 20 23

3.7   Derived variables

Sometimes observations are processed in order to generate a **derived number.** Examples of derived variables are *ratios, proportions, percentages* and *rates*.

A **ratio** is the simple relationship between two numbers measured on the same ratio scale. For examples, if the mid-arm circumference of a new born baby is 8.7 cm and the chest circumference is 29.8 cm, then the arm : chest circumference ratio is 8.7 : 29.8. Alternatively, the chest : arm circumference ratio is 29.8 : 8.7. Notice how the *units* of the scale disappear in the ratio – that is because they cancel out.

One value in a pair (usually the smallest) may be converted to unity by division. Records from a mobile blood donor unit showed that in donors the ratio of blood groups type AB to A was 72 : 720, that is, 1 : 10. A ratio may also be expressed as a fraction. For the blood group example it is 1/10, and when this is reduced to a decimal, 0.1, it is commonly referred to as a *coefficient*.

A **proportion** is the ratio of a *part* to a *whole*. When multiplied by 100, a proportion is converted to a **percentage**. If, in an eight hour shift, a worker spends two hours engaged upon paperwork, the proportion of the time spent is 2 : 8, i.e. 0.25. If a proportion is based upon counts of things it may be referred to as a *proportional frequency*, that is, the ratio of the number of individuals in a particular category to the total number in all categories. This may be illustrated by referring again to blood group types.

Example 3.1

Records from a blood donor unit show the total number of donors exhibiting four blood group types are: A, 725; B, 258; AB, 72; and O, 1073. What is the proportion (proportional frequency) of each blood group type in the sample?

The total number N of donors is $725 + 258 + 72 + 1073 = 2128$.

The proportion is give by:

$$p_i = \frac{n_i}{N}$$

where p_i is the proportion of a particular category, n_i is the number of individuals in a particular category and N is the total number in all categories. The proportion and percentage of each blood group type is shown in the following table:

Blood group type	n_i	n_i/N	p_i	Percentage
A	725	725/2128	0.34	34
B	258	258/2128	0.12	12
AB	72	72/2128	0.034	3.4
O	1073	1073/2128	0.50	50
	$N = 2128$		$\Sigma p_i = 1$	$\Sigma = 100$

Notice that the sum of the proportions equals 1 (Σ means 'sum of'), and the sum of the percentages equals 100 (allowing for small rounding errors).

Conventionally, a **rate** is the value of some variable *standardized* to a convenient unit of time, for example, breaths per minute. A rate may be standardized to additional variables, e.g. birth rate or death rate may be recorded as the number of births (or deaths) per year per 1000 of the population. This may be standardized yet further, for example, in the case of *age-specific rates*. Thus, the *infant mortality rate* is defined as the number of deaths during a calendar year among infants under one year age, divided by the total number of live births during that year. It is one of the most important measures of the health status of a nation.

Incidence rate refers to the reported number of infections per unit time, for example, the number of ear infections in 1000 junior schools in a two month period (see Section 6.2).

Statistical techniques may be applied to derived variables; sometimes the data have to be converted or *transformed* (see, for example, Section 9.11).

3.8   Logarithms

Logarithms are a special form of derived variable frequently used by scientists of all kinds. In practice, the logarithm ('log') of a number is obtained by simply entering the number in a scientific calculator and pressing the 'log' key. It may be converted back by 'anti-logging' it – on most calculators this is accomplished by pressing the 'INV' (inverse) key before the 'log' key. An advantage of using the logarithms of numbers is that they compress numbers which may span several orders of magnitude (say, 10 to ten million) onto a convenient scale. Thus, the log of 10 is 1 and the log of ten million is 7. See Section 9.11 for a practical example of how logarithms can 'squash up' a scale.

The most commonly used logarithm is 'log to the base 10' ($\log_{10}$). An alternative is the 'natural logarithm', or log to the base e, denoted by 'ln' or '$\log_e$'. This is obtained by entering the number and pressing the 'ln' key. Both sorts of logarithm exert the 'squashing up' effect, and sometimes the choice of which one to use is no more than arbitrary. If in doubt, use $\log_{10}$.

Thus, $\log_{10} 25 = 1.398$; $\ln 25 = 3.219$.

3.9   The precision of observations

When an observation is of a discrete variable, that is, a **count**, we are usually sure of its precision. There may be *exactly* four donors with blood group type B who present themselves in a session at a blood donor unit. A measurement, on the other hand, is never exact; it is precise only to within certain limits.

If the distance a trolley has to be pushed from the A & E reception to the radiography unit is measured with a tape marked in 1-metre intervals, measurements are precise to the nearest whole metre. An observation can be recorded as 10 ± 0.5 m. This implies that all distances between the limits of 9.5 m and 10.5 m are recorded as 10 m, as shown in Figure 3.1.

For smaller lengths we could use a more finely graduated scale, for example a tape marked in 10 cm (0.1 m) divisions. Each observation is then precise to the nearest 0.1 m and an observation of 10.6 m is written

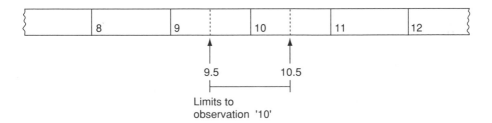

Figure 3.1   The limits of an observation

as 10.6 ± 0.05 m; that is, all distances between 10.55 m and 10.65 m are recorded as 10.6 m.

We could continue increasing the precision of measurements by using a metre rule graduated in millimetres, Vernier callipers capable of recording to 0.01 mm, or a microscope eye-piece graticule down to 0.001 mm. In each case, the measurement is precise to *plus or minus half the interval spanned by the last measured digit*; in the case of the graticule measurement this is $\pm$ 0.0005 mm. An observation of 1.364 mm is within an interval spanning 1.3635 mm to 1.3645 mm.

Note the distinction between *precision* and *accuracy*. An expensive spring balance might be precise, weighing to 0.1 g, but if it is badly adjusted it will not be accurate. A broken clock is accurate twice a day!

3.10   How precise should we be?

Since it is clearly possible to choose (within reason) the degree of precision of a measurement, the question arises, 'how precise should we make our measurements?' We should not choose, for example, to measure the distance between two departments in a hospital in millimetres.

A commonly accepted principle is that an observer should aim to have between 30 and 300 unit steps between the largest and smallest observations. For example, a series of measurements that ranges from 59 mm to 67 mm has only $(67 - 59) = 8$ unit steps, well below the suggested lower limit of 30. An error of one unit in eight represents an unacceptable 12.5 percent. Measurement to a tenth of a millimetre would give about 80 steps. We do not know exactly how many steps because the largest length could be, say, 67.4 and the smallest 58.8, a difference of 86 unit steps. This is more satisfactory – an error of 1 unit in 80 is an acceptable 1.25 percent.

3.11   The frequency table

Collected data should be organized and summarized in a form that allows further interpretation and analysis. Table 3.1 shows the weights (to the nearest 0.1 of a kilogram) of 100 new-born babies.

The weights are presented in the same order as the babies were weighed, and are therefore *ungrouped*. A quick scan of Table 3.1 reveals

Table 3.1 Weights of 100 babies (kg)

3.2	3.0	2.9	2.8	3.1	3.1	3.1	3.1	3.1	3.4
2.9	3.4	2.5	3.0	3.5	3.1	3.0	3.2	3.6	3.3
3.1	3.3	3.1	3.4	2.9	3.2	2.8	3.3	3.2	2.8
2.9	2.6	3.0	3.1	3.1	3.0	3.0	2.9	2.9	3.1
2.8	3.1	3.3	3.0	2.8	3.1	3.5	3.2	3.3	2.8
3.0	3.3	3.1	2.7	3.0	3.2	3.6	2.9	2.9	3.4
3.3	3.1	3.1	3.2	2.9	3.0	3.4	2.7	3.2	3.5
3,4	3.2	3.0	2.6	3.3	3.5	3.1	2.8	3.4	3.0
2.7	3.0	3.2	3.2	2.9	3.1	2.7	3.1	3.0	3.3
3.3	3.0	2.9	3.2	3.3	3.2	3.7	3.2	3.1	3.2

that particular values are repeated a number of times: there are, for example, five values of 3.1 kg in the top row alone. The value of *3.1 kg* is called a **frequency class** and, rather than record all 100 values individually (as in the table), it is more economical of space, and more revealing, to *group* the data into all the frequency classes. We should remember that each frequency class is a **class interval** with implied limits of ± 0.05 kg. Thus, all weights between 3.05 and 3.14 are placed in the '3.1 kg' class. The grouped observations are shown in Table 3.2.

The data in Table 3.2 are grouped into columns: implied class interval, frequency class x and f. The *tallies* are presented in this example to give a visual appreciation of how the frequencies are distributed between the frequency classes. The two columns x and f represent a **frequency**

Table 3.2 Grouped new-born baby weights

Implied class interval	Frequency class x (kg)	Tallies	Frequency f
2.45–2.54	2.5	1	1
2.55–2.64	2.6	11	2
2.65–2.74	2.7	1111	4
2.75–2.84	2.8	1111111	7
2.85–2.94	2.9	11111111111	11
2.95–3.04	3.0	111111111111111	15
3.05–3.14	3.1	11111111111111111111	20
3.15–3.24	3.2	111111111111111	15
3.25–3.34	3.3	11111111111	11
3.35–3.44	3.4	1111111	7
3.45–3.54	3.5	1111	4
3.55–3.64	3.6	11	2
3.65–3.74	3.7	1	1

$$n = \Sigma f = 100$$

table. Although this particular table has been constructed for interval measurements (weights), frequency tables can also be constructed for count data and for nominal and ordinal scales. The manner in which frequencies are distributed between the frequency classes is described as a **frequency distribution**.

Readers will note that there are only 12 unit steps between the largest and smallest observations in Table 3.2, and an increase in precision is highly desirable. The purpose of the 'rounded' data is simply to illustrate the construction of a frequency table.

3.12   Aggregating frequency classes

When the spread of observations is large and the number of observations relatively few, then a frequency distribution appears drawn out and disjointed if every step interval corresponds to a frequency class. In such cases it is advisable to aggregate, or group, adjacent classes to smooth out the distribution. Example 3.2 shows how to do this.

Example 3.2

A sample of 50 measurements of systolic blood pressure of male patients following myocardial infarct is presented in Table 3.3. We note first that there are $(200 - 156) = 44$ unit steps between the largest and smallest observation; the data are sufficiently precise. Secondly, each observation is measured to the nearest whole unit and is therefore precise to within $\pm$ 0.5 units. The first observation in the table (162) therefore falls within an implied interval of 161.5 to 162.5 units.

The data are summarized and grouped into the frequency table in Table 3.4.

Table 3.3   Systolic blood pressure of 50 males after myocardial infarct

162	188	173	168	174	183	167	186	177	187
170	174	164	174	159	177	173	163	180	196
171	156	184	179	190	181	166	181	182	176
169	172	174	162	175	192	178	177	200	191
188	168	165	179	193	175	160	180	187	176

Table 3.4 A grouped frequency table

Implied limits of each class interval	Class mark (mid-point of class)	Tallies	Frequency f
155.5–158.5	157	1	1
158.5–161.5	160	11	2
161.5–164.5	163	1111	4
164.5–167.5	166	111	3
167.5–170.5	169	1111	4
170.5–173.5	172	1111	4
173.5–176.5	175	11111111	8
176.5–179.5	178	111111	6
179.5–182.5	181	11111	5
182.5–185.5	184	11	2
185.5–188.5	187	11111	5
188.5–191.5	190	11	2
191.5–194.5	193	11	2
194.5–197.5	196	1	1
197.5–200.5	199	1	1
			$n = \Sigma f = 50$

The steps involved in the construction of Table 3.4 are listed below. They enable the construction of frequency tables from any set of measurement data with a degree of class aggregation that suits specific needs.

1. Determine the *range* of scores: the highest observation minus the lowest observation plus 1. (One is added to take into account the implied limits of the numbers.)

$$\text{Range} = \text{highest observation}(200) - \text{lowest observation}(156) + 1$$
$$= (200 - 156) + 1 = 45$$

Decide how many categories (class intervals) are required. Normally the number of class intervals is not less than 10 and not more than 20. In this case, we have selected 15 as a convenient number of classes.

2. Divide the range by the number of class intervals required. This gives the number of unit steps that are aggregated to make up an interval class.

Number of unit steps per class interval $= 45 \div 15 = 3$

If the calculated number of unit steps per class interval is a fraction, then round to the nearest whole number.

3. Construct the class interval column starting at the top with the lower limit of the smallest observation (155.5). Add the class interval size (3) to this lower limit. The range of the lowest class interval becomes 155.5 to 158.5. The lower limit of the next class becomes 158.5, to which 3 is added to give the class interval range 158.5 to 161.5. This procedure is repeated, moving down the column until the class interval column includes an interval into which the largest observation (200) can be placed, namely, 197.5 to 200.5.

4. In the next column insert the mid-point of the class. This labels the class and is called the **class mark**. It is obtained by adding half the class interval size to the lower limit of the class: thus, $155.5 + 3/2 = 157$.

5. In the next column provided, insert a tally for each individual observation in the raw data table. For example, for the observation 162, a tally is inserted to show that it falls into the class interval range 161.5 to 164.5.

6. Total up the tallies within each class interval and place in the frequency column in line with the appropriate class.

7. Total the frequency column (n). This serves as a useful check that all data have been included in the table.

3.13   Frequency distribution of count observations

In principle, the construction of a frequency distribution of count observations is similar to that for measurement observations. However, because each observation, and hence, frequency class, has an *exact* value (there is no implied upper and lower limit to a count of 150 babies), the column of *implied class interval* is redundant.

A minor problem may arise if the frequency classes of count data are aggregated. Imagine that the observations in Section 3.11 are counts, for example, of the number of patients admitted to 50 hospitals over a standardized time period. The construction of a frequency table is undertaken exactly as described, except that the column of implied limits is redundant. The class mark (the mid-point of each class interval) is still a useful number, however.

Suppose that we had decided to aggregate the unit steps into classes of four instead of three. In Step 4 of our instructions, the range of the lowest class interval would be 155.5 to 159.5. To find the class mark we would add half of the class interval range to the lower limit, that is, $155.5 + 4/2 = 157.5$. A class mark that is a *fraction* is not sensible in a frequency distribution of counts; a frequency class labelled '157.5 patients' has little meaning. To avoid this anomaly when constructing a frequency distribution of count observations in which the classes are aggregated, make sure that the number of units steps in each class interval is an *odd number*. The class mark is then a whole number.

.14 Bivariate data

It often happens that we obtain more than one observation from a unit in a sample. A child may have its weight *and* height measured; a patient may have the pulse rate *and* breathing rate measured. A set of observations of *two* variables from each item or unit in a sample is called **bivariate data**. There are statistical methods available for analysing bivariate data that we shall explain in later chapters.

4 PRESENTING DATA

4.1 Introduction

One drawback in presenting data in the form of a frequency distribution (Table 3.2) is that the information contained there does not become immediately apparent unless the table is studied in detail. To simplify the interpretation of the information, and to pin-point patterns and trends, the data are often processed further and transformed into a visual presentation. The most common methods of presenting data are based upon graphical techniques. In this section we describe methods suitable for presenting data that nurses and health care workers are likely to encounter.

4.2 Dot plot or line plot

The dot plot is a method of presenting data that gives a rough but rapid visual appreciation of the way in which data are distributed. It consists of a horizontal line marked out with divisions of the scale on which the variable is measured. A dot representing each observation is placed at the appropriate point on the scale. If certain observations are repeated, simply stack the dots on top of each other. Figure 4.1 shows two dot plots, each involving 16 observations (sampling units):

- Figure 4.1(a): weight of 17 new-born babies measured to the nearest 0.1 kg: observations are scattered about 3.1 kg.
- Figure 4.1(b): Numbers of fatalities resulting from road accidents received by 16 accident and emergency departments in a region during a weekend: 7 departments received no fatalities; 3 had one fatality; 2 had two, and so on.

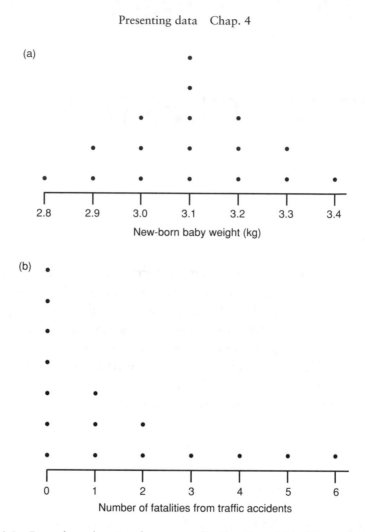

Figure 4.1 Dot plots showing frequency distribution of (a) 17 new-born baby weights; (b) numbers of traffic accident fatalities received by 16 accident and emergency departments

4.3   Bar graph

Portraying information by means of a bar graph is particularly useful when displaying data gathered from discrete variables that are measured on a nominal scale. Typically, a bar graph uses vertical lines (i.e. bars) to represent discrete categories of data, the length of the line being proportional to the frequencies within that category.

Suppose 31 donors offer themselves at a blood donor unit. After an initial test for blood group type, 15 are identified as type O, 10 of type A, 4 of type B and 2 of type AB. A frequency table may be constructed:

Blood group type	f
O	15
A	10
B	4
AB	2
	$n = 31$

Using these data, a bar graph may be constructed, as shown in Figure 4.2.

In its final form, the horizontal dashed lines are omitted; they are included here to show that the height of the bar corresponds to the respective frequency.

When observations are counts of things the bar graph is a useful way to present a frequency distribution. Illustrators often replace each bar with a vertical rectangle, or block, whose adjacent sides are touching. The frequency distribution of traffic accident fatalities shown as a dot plot in Figure 4.1(b) is shown as a bar graph in Figure 4.3, where the height of each block is still proportional to the frequency in each

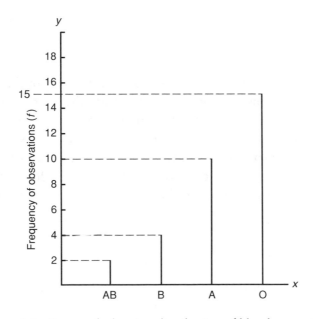

Figure 4.2 Bar graph showing distribution of blood group types

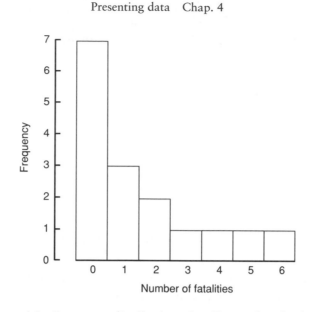

Figure 4.3   Frequency distribution of traffic accident fatalities

category because the width of each block is equal. When presented in this form the diagram is usually referred to as a **histogram**. Histograms are especially useful for presenting frequency distributions of observations measured on continuous variables, as we show in Section 4.4.

4.4   Histogram

The histogram is especially useful for presenting distributions of observations of **continuous variables**. In a histogram the *area* of each block is proportional to the frequency. The area of a single histogram block is found by multiplying the width of the block (the class interval) by the height (frequency). Almost invariably, the horizontal scale is marked in equal intervals, and then the blocks are of equal width and the height of each block is proportional to the frequency of observations.

Example 4.1

Observations of systolic blood pressure (mmHg) are obtained from 150 female hospital patients. The observations are shown in the form of a frequency table:

Systolic blood pressure (mmHg)	Number of patients (frequency)
100–109	7
110–119	16
120–129	19
130–139	31
140–149	41
150–159	23
160–169	10
170–179	3

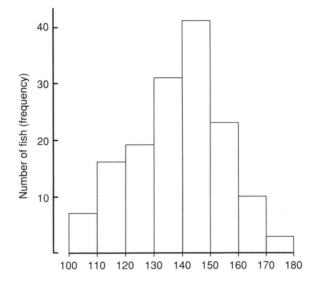

Figure 4.4 Frequency distribution of systolic blood pressure observations from 150 patients

The frequency distribution is presented in a histogram in Figure 4.4. Because the width of each block (class interval) is the same, 10 mmHg, the height of each block is proportional to the frequency. In Figure 4.4 the area of the first block is $10 \times 7 = 70$ units, compared with the fifth block which is $10 \times 41 = 410$ units. The area represented by the last three blocks is $(10 \times 23) + (10 \times 10) + (10 \times 3) = 360$ units.

4.5   Frequency polygon and frequency curve

If the mid-point of the top of each block in a histogram is joined by a straight line, a **frequency polygon** is produced. Figure 4.4 is reproduced

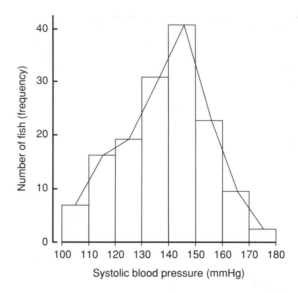

Figure 4.5   Frequency polygon of systolic blood pressure observations from 150 patients

with a frequency polygon superimposed in Figure 4.5. When the number of observations of a continuous variable is large and the unit increments are small, the 'steps' in the histogram tend towards a smooth, continuous curve, called a **frequency curve**. A frequency curve is superimposed

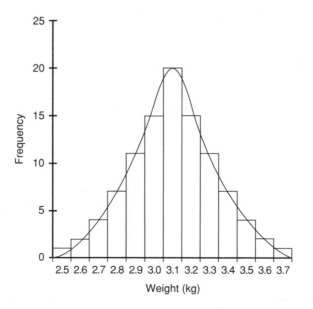

Figure 4.6   Frequency curve of 100 new-born baby weights

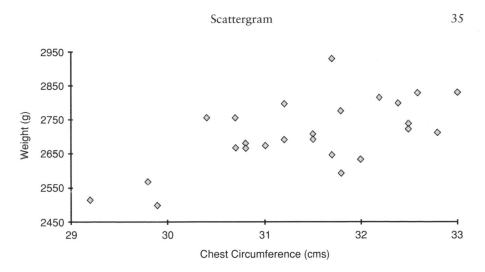

Figure 4.7   Scattergram of weight and chest circumference of 25 new-born babies

(in Figure 4.6) upon the distribution of 100 new-born baby weights given in Section 3.10.

4.6   Scattergram

When *pairs* of observations of two variables are obtained from each unit in a sample (that is, the data are bivariate), a scattergram (scatter plot) is used to display the data. In Figure 4.7, each point in the scattergram represents an individual baby for which weight *and* chest circumference are indicated. In two-dimensional presentations like this, it is conventional to refer to the vertical scale as the *y*-axis and the horizontal scale as the *x*-axis.

4.7   Circle or pie graph

The pie graph is best suited for displaying data that are *percentages* or *proportions*. If the area of a circle is regarded as 100 percent it can be divided into sectors (the 'slices of the pie') that correspond in size to each individual percentage or proportion making up the total. To work out the angle of each sector of the pie divide each percentage by 100 and multiply by 360, the number of degrees in a circle. Proportions are simply multiplied by 360. However, with the readily availability of graphical programmes it should rarely necessary to draw a graph 'by hand'.

Table 4.1   Gender differences in locations of accidents (percentages in brackets)

Gender	Location of accident						
	Home & garden	Road	Workplace	School or college	Sports area	Other	Total
Males	531 (27.2)	322 (16.8)	512 (26.7)	152 (7.9)	190 (9.9)	208 (11.4)	1915
Females	653 (48.5)	272 (20.2)	136 (10.1)	82 (6.1)	41 (3.0)	163 (12.1)	1347

Example 4.2

The numbers of accidents in different locations for males and females are shown in Table 4.1, together with the percentages in brackets.

Pie graphs of these data are shown in Figure 4.8.

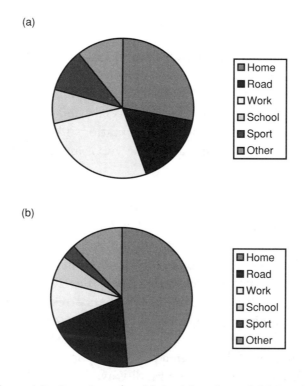

Figure 4.8   Location of accidents (a) males and (b) females

When the number of categories in a pie graph is large, or when some of the slices are very narrow, pie graphs can become cluttered and difficult to interpret. Then, it is worth considering the use of a bar graph, as it is easier to rank categories in a bar graph and it does not depend on the ability to judge angles correctly.

5 CLINICAL TRIALS

5.1 Introduction

The survey methods we describe in Chapter 2 provide information on the current state of affairs, and can be used, for example, to assess how well inhalers are being used in an asthma clinic; to elicit patient opinions; or to quantify waiting times in an outpatient clinic. However, we need a different approach to assess the effect of an intervention, for example, changing from a standard treatment to a new one; or the effect of moving hospital outpatient care to general practice clinics; or evaluating the efficacy of an established approach to health care. Health care investigations in which the specific objective is compare two or more treatments, and if possible identify which is the most effective, are known as **clinical trials**. The essential feature of a clinical trial is the use of randomized allocation to assign individuals to the *treatment* and *control* groups. The only difference between the treatment and control groups is the treatment. In all other respects, the groups are as similar as possible. Randomization is used to safeguard against selection bias, and is insurance (in the long run) against accidental bias, and is the underlying principle of statistical testing. We emphasize that the subject of clinical trials is a large one, and in this introductory text we do no more than offer a brief outline. Readers interested in developing their knowledge in this area are referred to Pocock (1983).

5.2 The nature of clinical trials

The key to good clinical trials has three stages:

1. Formulate a hypothesis to realize the study aim.
2. Design an experiment to assess the validity of this hypothesis.
3. Perform the study, collate the results and draw conclusions.

To design the experiment the hypothesis should be specific, and not be a vague statement like: 'drug A is better than Drug B', since this does not define either 'how much is better' or even what we mean by 'better'. Thus, Drug A might cure 100 per cent of patients, but has nasty side-effects while Drug B cures only 50 per cent and has no side-effects.

The size of the difference in treatment effect is also very important. For example, an anti-hypertensive treatment that lowers systolic blood pressure by 20 mmHg is far more impressive than one that lowers systolic blood pressure by 5 mmHg.

5.3   Clinical trial designs

There are three main clinical study designs: *group comparative, cross-over* and *sequential*, the main features of which are summarized in Table 5.1.

In **group comparative studies**, patients are first monitored and, at the end of this baseline period, eligible patients are randomly allocated to one of the treatment groups. The trial is finished and the collected data are analysed once an agreed number of patients have completed the study. Details of how we choose the number of patients in each group can be found Appendix 12.

In a **cross-over study**, each eligible patient receives all treatments in a pre-specified order, usually with a 'washout' phase between each of the treatment periods. For example in a 2×2 (two treatments and two treatment periods) cross-over study, half the patients are allocated to a 'treatment order' group which receives Treatment A in the first treatment period, and are then 'crossed over' to Treatment B for the second treatment period. The other 'treatment order' group receives Treatment B in

Table 5.1   Features of clinical trial designs

Feature	Design type		
	Group Comparative	Cross-Over	Sequential
Treatments received by each patient	one	all	one
Fixed treatment group sizes	yes	yes	no
Number of analyses	one	one	many

Table 5.2 Two treatment, two treatment periods (2 × 2) cross-over design

Treatment order	Treatment period one	Treatment period two
AB	A	B
BA	B	A

the first treatment period and are then 'crossed over' to Treatment A for the second treatment period (Table 5.2).

The advantage of the cross-over design is that each patient receives both treatments, and we therefore obtain a 'within-patient' comparison of treatments. This greatly reduces the number of required patients in the study. However, there are two drawbacks. First, for the within-patient comparison to be valid, each patient must return to a similar clinical state at the beginning of the second treatment period as they were in at the beginning of the first treatment period. Thus, cross-over designs are only suitable for the assessment of chronic disorders in which we compare alleviation of symptoms (asthma, for example). Even here, we have to be careful if the treatment periods are long, say three months, as the results may be confounded by seasonal differences. Secondly, the studies are longer because we have a number of treatment periods rather than just one. Further details of cross-over designs are described by Jones and Kenward (1989).

In a **sequential study** design, patients are allocated to a single treatment group and the data are analysed at pre-agreed intervals (for example, after every ten patients have completed the study). The study is deemed complete when one treatment is found to be better than the other. This differs from the group comparative and cross-over designs in which the data are only analysed once, when a pre-specified number of patients have completed the study. Further details of sequential study designs are provided by Whitehead (1982).

5.4 Psychological effects and blind trials

The factors we have to consider when undertaking clinical trials include **psychological effects**. If patients believe that they are being treated, they can get better, or even, in some cases, experience adverse events. These effects may be seen both when a standard treatment is being used as the 'control', or when the control group is being treated with a **placebo**. A placebo treatment is identical to the active treatment group (for example, the same appearance, taste, smell, labelling, treatment regimen) differing only in the absence of the pharmacologically active agent.

Also, there is a risk that patients and investigators are subjective in their assessment, which could be affected by knowledge of treatment. Thus, an investigator might introduce bias by anticipating a particular outcome (perhaps subconsciously) from knowing previous results. This particular problem is overcome through the use of **blind trials**. In a *single blind study*, the patient does not know the treatment they are taking. In a *double-blind study*, neither the patient *nor* the investigator know which treatment the patient is taking. In any study, the level of 'blinding' depends on the nature of the intervention being studied. However, we suggest that it is possible (and desirable) in all cases to implement 'blinding' at some level. Even if we are not able to 'blind' the patient or investigator to the intervention (for example, a surgical procedure), the person performing the assessment of intervention effect does not need to know details of the intervention the patient received.

The majority of definitive clinical trials tend to be **randomized double-blind group comparative studies**, and we endorse their use wherever possible.

5.5   Historical controls

The results of clinical trials are more likely to be reliable if the 'treatment' and 'control' observations are made concurrently, that is, the two groups are studied at the same time. However, it is sometimes possible to use historical data. In this situation, suitable patients receive the 'new' treatment and their outcome measures are compared with those obtained from the records of patients previously given the former treatment. The use of historical controls pre-supposes that all variables (for example, the type of patient, the severity of the illness, the ancillary treatments given, the measurements made) except the new treatment have remained constant in time. All factors must remain the same for the previously treated patients as for those receiving the new treatment. However, we warn that past observations are unlikely to relate precisely to a similar group of patients The quality of the information extracted from historical controls is likely to be inferior, since former patients were not initially intended to be in a trial. Patients being given a new, and as yet unproven, treatment are more likely to be closely monitored and receive more intensive ancillary care than the historical controls that received the orthodox treatment and were not part of the study. For these, and a variety of other reasons, studies with historical controls are much more likely to over-estimate the efficacy of a new treatment. We suggest that historical controls in clinical trials are of limited value, and should be avoided whenever possible.

5.6 Ethical issues

In conducting clinical trials there are ethical issues to take into account. Thus, it is unethical for an investigator not to give patients the best possible treatment. In a study of two treatments, half the patients will, by definition, receive an inferior treatment. If the investigators are not sure which is the best treatment for a patient, then they themselves should participate in a study.

Moreover, it is unethical not to discover whether a new treatment is an improvement, since this would deny future patients the possibility of a cure. It is also unethical to perform bad trials that give misleading results, and thereby encourage others not to use a treatment that is beneficial, or to use a treatment that is not beneficial, or may indeed have harmful side-effects. The converse of this is that good clinical trials, as outlined in Section 5.2, that have clearly defined aims, adequate sample sizes, and are carefully planned and monitored, are ethical.

5.7 Case study: Leicestershire Electroconvulsive Therapy (*ECT*) study

Prior to this study, ECT had been in use for over 50 years as a treatment for severe depressive illness. ECT involves inducing a convulsion by passing an electric current through the brain. The convulsion is modified by giving a muscle relaxant and an anaesthetic, so that only a few muscle twitches are produced. In 1977 the Royal College of Psychiatrists claimed that 'Evidence in favour of ECT for depressive states is incontrovertible'.

However, no hard evidence was available to support this statement and, at the time, no clinical trials had ever been performed.

In 1980, the Medical Research Council (MRC) funded a randomized double-blind group comparative study at Northwick Park which concluded (Johnstone *et al.* 1980) that:

- ECT had only a small effect at the end of the trial period.

- No difference in the condition of patients given real and simulated treatment at one- and six-month follow up visits.

The decision was made in Leicestershire to run a further study to assess the effectiveness of ECT. A brief summary of the study is given in Table 5.3. Further details are available in Brandon *et al.* (1984).

Table 5.3　　1981/2 Leicestershire Electroconvulsive Therapy study

Aim	To assess whether ECT is effective and, in particular, is it effective for patients with depressive illness.
Treatments	Eight shocks (either real or simulated) usually given twice weekly for four weeks. (Procedure for simulated ECT exactly the same as for real ECT without the final shock.)
Design	Double-blind group comparative study. Following a baseline period, eligible patients were allocated to either the real ECT or simulated ECT treatment groups. During the follow up period, the consultants could give whatever treatments they considered appropriate.
Assessments	Week 0: Baseline Week 2: Half-way through four week treatment period Week 4: End of four week treatment period Week 12: Follow-up visit
Study population	All patients for whom ECT therapy was prescribed from a total population of 840 000 within Leicestershire.
Measurements	Numerous rating scales including: 　　Hamilton Depression Rating scale (HDR) 　　Montgomery and Asberg Schizophrenia scale (MASS) 　　Visual Analogue Depression scale (VA1) 　　Visual Analogue Global scale (VA2) All these were interviewer assessments. For each scale, higher values indicate a worse condition.
Primary Endpoint	Change in HDR scale at the end of the four week treatment period.

The data from the subset of patients with depressive illness are given in Appendix II.

There are two questions to consider when analysing the results of this study. The first considers whether the 'real' ECT and 'sham' ECT treatment groups are comparable at the beginning of the study. Randomization was used to allocate patients to receive either real ECT or sham ECT. Unfortunately, the use of randomization does not guarantee that this will be the case, but it does give a better chance than any other method that the groups are comparable. It is usual to check baseline comparability using descriptive statistics, similar to those described in Chapters 7 and 8, and graphical summaries rather than formal statistical tests. If treatment groups differ in some systematic way, the study results are more difficult to interpret.

The second question considers the *objective* of the study, namely, is there any difference between the treatments? To facilitate the appreciation of this, we calculate **changes** from the baseline scores for each subject in the study. This procedure removes *between-subject variation* from the comparison of treatments, and uses the changes from the baseline to compare the real ECT and sham ECT treated groups. If the changes from baseline are similar for the two groups, this indicates that the two treatments have similar effects. If one group has improved more than the other, this indicates that the treatment of that group is more beneficial than the other. Details of how to compare two treatment groups are presented later in Chapter 15.

How should we measure the **change** from the baseline? There are a number of possibilities. For example, absolute differences or percentage changes are possible. An ideal measure of change is at least correlated with the baseline values. One way to check this is to use a scatter plot, plot the change measure in the vertical axis against the baseline measurement on the horizontal axis. A plot showing a random scatter of points indicates no correlation. The measurement of correlation is described in Chapter 13.

5.8 Summary

We have introduced several ideas in this chapter. The main ones which define a 'clinical trial' are (a) the study objective is to compare two or more 'treatments', and (b) that randomization is used to allocate eligible patients to the treatment group.

We refer to the Leicestershire ECT study data in Appendix 11 to illustrate examples in other sections of this book, for example, in Section 15.11 and Appendix 12.

6 INTRODUCTION TO EPIDEMIOLOGY

6.1 Introduction

In this chapter we provide a brief introduction to epidemiology and epidemiological studies. A more advanced treatment of the subject can be found in Ahlbom and Norell (1984).

Epidemiology is not an easy term to describe briefly; however, here are a few of our favourites definitions:

'Who gets what, why, when, where and what happens.'

'The study of the distribution and causes of health impairment in human populations and the evaluation of actions taken to improve health.'

'The study of epidemic diseases with a view to finding means of control and future prevention. This not only applies to the study of such classical epidemics as plague, smallpox and cholera but also includes all forms of disease that relate to the environment and ways of life. It thus includes the study of links between smoking and cancer, and diet and coronary diseases, as well as communicable diseases.'

Studies in epidemiology are used in a variety of ways. For example, they are used to identify health problems in communities; to identify disease determinants (that is, to identify factors which make a person at greater or lesser risk of developing a disease); to develop strategies for intervention; and to evaluate the effectiveness of intervention programmes

Epidemiological studies investigate characteristics of human populations in the form of an experiment in which groups for comparison are selected on the basis of exposure to a particular risk factor. For example, to investigate the relationship between lung cancer and smoking, a group of 'smokers' could be compared with a group of 'non-smokers'. This example highlights the fundamental difference between epidemiological studies and clinical trials (Chapter 5). Thus, in a clinical trial, we use randomization to allocate individuals to the 'treatment' and 'control' groups, whereas in epidemiological studies the groups are 'self-selected'. We may decide to choose the individuals to participate in the epidemiological study randomly, but each of the selected individuals has been exposed to the risk factor under investigation either by choice (for example, smoking) or inadvertently (for example, living close to high voltage power cables).

In epidemiology the aim is to identify factors associated with individuals being at either a greater or lesser risk of developing a disease, and to try to understand how disease occurrence is related to different characteristics of individuals or their environments.

6.2   Measuring disease

The primary measures of disease are **prevalence** and **incidence**. Prevalence describes the proportion of the population that has the disease in question at one specific moment in time:

$$\text{Prevalence} = \frac{\text{number of individuals having the disease at a specific time}}{\text{number of individuals in the population at that point in time}}$$

Example 6.1

In a study of eyesight and the occurrence of a certain eye disease, a sample of 2400 people aged between 50 and 85 were examined, and 300 were found to have cataracts:

$$\text{Prevalence of cataracts within this sample} = \frac{300}{2400} = 0.125$$

Like all proportions, prevalence is a dimensionless quantity taking values between 0 and 1. Prevalence is like a 'snap-shot' of the population at a particular point in time, and does not take into account the duration of time that the individuals have had the disease.

The incidence rate describes the frequency of new cases that occur during a time period, and is the basic measure of disease occurrence:

$$\text{Incidence} = \frac{\text{number of cases of the disease that occur in a population during a period of time}}{\text{sum for each individual in the population of the length of time at risk of getting the disease}}$$

Example 6.2

The data in Table 6.1 were collected during a study to investigate the possible association between the use of Oral Contraceptives (OC) and the development of breast cancer. A sample of disease-free women were classified according to OC status ('current user', 'never user'). The women were assessed every two years, and their OC use and breast cancer status were recorded. The length of time that each woman had used or never used OCs was calculated, and these 'person-times' accumulated over the entire sample.

Incidence of breast cancer amongst 'current users' $= \frac{9}{2935} = 0.00307$ events per person year or (multiplying by 100 000) 307 events per 100 000 person-years.

Incidence among 'never users' $= \frac{239}{135130} = 0.00177$ events per person-year or 177 events per 100 000 person years.

When is it preferable to report 'incidence' rather than 'prevalence? To answer this question it is helpful to consider individuals to be in one of two 'states': a *disease-free state* or a *disease state*. In epidemiological studies where the main aim is to explore causal theories, or to evaluate effects of preventative means, the interest is focused on the flow of cases from the disease-free state to the disease state and therefore incidence is the best measure. However, when planning health service resource

Table 6.1 Relationship between breast cancer and oral contraceptive usage

OC-use group	Number of cases of breast cancer	Number of person-years
Current users	9	2935
Never users	239	135 130

requirements, or assessing the need for medical care in a population, prevalence is considered to be more useful.

Studies on chronic diseases, such as rheumatoid arthritis or diabetes, generally employ prevalence measures while studies in acute situations, like cancer or myocardial infarction, generally use incidence measures.

6.3 Study designs – cohort studies

The two designs most often used to identify whether a risk factor is associated with disease are **cohort studies** and **case-control studies**. In cohort studies (Figure 6.1), groups of exposed and unexposed individuals (the **cohorts**) are identified and followed for a period of time. In our smoking example, the 'exposed' cohort comprise smokers, and the 'unexposed' comprise non-smokers. Records are made of whether or not subjects develop the disease under investigation. If the disease under consideration is lung cancer, we might expect by chance that some individuals from the exposed group do *not* develop the disease while, conversely, some from the unexposed group *will* develop the disease (represented by the diagonal arrows in Figure 6.1). We are of course looking for significant differences in the proportions of disease development between the exposed and unexposed cohorts. Note that in cohort studies all individuals who have the disease at the start of the study (e.g. already have lung cancer) are excluded.

Example 6.3

The data in Table 6.2 were collected in a study to investigate the association between smoking and mortality from coronary heart disease.

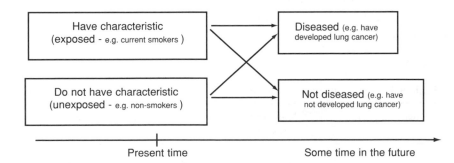

Figure 6.1 The principle of cohort studies

Table 6.2 Mortality from coronary disease among smokers and non-smokers

| Age | Number of deaths per 100 000 person years | | $I_{exposed}/I_{unexposed}$ (Relative risk) |
	Smokers (Exposed)$I_{exposed}$	Non-smokers (Unexposed)$I_{unexposed}$	
35–44	60	10	6.0
45–54	240	110	2.2
55–64	720	490	1.5
65–74	1470	1080	1.4
75–84	1920	2180	0.9
All ages	440	260	1.7

To calculate the age-specific incidence rates, we divide the number of deaths from coronary heart disease by the number of person-years. In the 35–44 age group, for example, the incidence of mortality from coronary heart disease is 60 per 100 000 person-years for smokers compared with 10 per 100 000 person-years for non-smokers.

The ratio of incidence rate in the exposed group to that of the unexposed group is used to quantify the association between exposure and disease occurrence. This comparison of incidence rates is known as **relative risk** or **rate ratio**. For example, smokers in the 35–44 age-group have a relative risk of 6; that is, they are six times more likely to suffer coronary disease than non-smokers.

One difficulty associated with cohort studies is the need to identify cohorts of exposed and unexposed individuals, and then to monitor both groups for a period of time. The circumstances when this problem is greatest are during studies of rare diseases, or diseases that exhibit a long time interval between exposure and occurrence. Then, **case-control studies**, in which data are collected retrospectively rather than 'prospectively', are appropriate. In these studies, data are collected retrospectively, rather than by conducting new trials.

6.4 Study designs – case-control studies

The principle of **case-control studies** is illustrated in Figure 6.2. This example relates to whether patients who have had a history of blood transfusion are more at risk of developing Hepatitis C. First, individuals are identified who currently have the disease (the 'cases'), and also a control group of individuals without the disease. Then, the history of

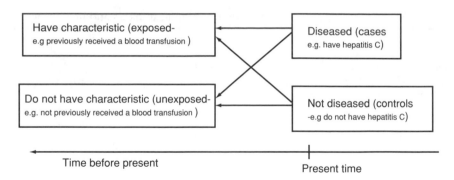

Figure 6.2 Principle of case-control studies

occurrence of the specific exposure (previous blood transfusion or not) is collected for each person in the study. Finally, the proportion of 'exposed' individuals in the 'case' group to 'unexposed' individuals in both the control group *and* the case group are compared. If the proportion of exposed to unexposed individuals among cases was greater than that of controls this offers support to the hypothesis that exposure is related to disease. However, if the proportion of exposed to unexposed individuals is similar for both 'case' and 'controls', this raises doubt about the association between exposure and disease.

Example 6.4

The data below were collected in a case-control study investigating the association between smoking and cancer of the mouth and pharynx:

	Smokers		
	Yes	No	All
Cases	457	26	483
controls	362	85	447

The ratio of unexposed (non-smokers) to exposed (smokers) is 457/26 for 'cases' and 362/85 for 'controls' approximates to the true incidence rates (remember, to calculate the true incidence rate we divide the number of cases by the cumulative 'person-years exposure'; here, we don't know the actual person-years exposure it is assumed to be similar for both groups). The relative risk is the ratio of the incidence rates among exposed to unexposed; it is approximately equal to the ratio of unexposed to exposed among cases and controls, that is:

$$\text{Relative Risk} = \frac{\text{Incidence}_{cases}}{\text{Incidence}_{controls}} \approx \frac{457/26}{362/85} = \frac{457 \times 85}{26 \times 362} = 4.1$$

The incidence rate is about four times greater for smokers than for non-smokers, suggesting that the risk of developing the cancer is about four times higher in smokers. At which point the risk is considered to be 'statistically significant', rather the result of a chance effect is decided by a statistical test. Suitable tests for comparing proportional frequencies are chi-square tests, the subject of Chapter 12 (see Example 12.7).

An alternative way to present the relative risk is the inverse of the above equation, which amounts to a measure of the *reduction* in risk of cancer in non-smokers as compared to smokers:

$$\text{Relative Risk} = \frac{\text{Incidence}_{controls}}{\text{Incidence}_{cases}} \approx \frac{362/85}{427/26} = \frac{26 \times 362}{457 \times 85} = 0.242$$

Thus, the risk to non-smokers is 0.242 (= 1/4.1), in other words, about a quarter of risk to smokers. However, we consider that the risk is better understood when expressed in terms of *increased* risk. For example, the message that smokers are at four times the risk of cancer of the mouth and pharynx than non-smokers is likely to be more easily understood than one indicating that non-smokers are at quarter of the risk than smokers.

A relative risk close to 1.0 offers no support to the hypothesis that exposure and disease are associated. How much larger than 1.0 the risk factor needs to be depends on the sample size. As indicated above, a statistical test is required to decide objectively how much greater than 1.0 is 'significant'.

6.5   Cohort or case-control study?

Cohort studies have an established successful reputation in epidemiology. However, they can be subject to bias (including diagnostic bias), and also to loss of subjects during follow-up. They are also expensive and extremely difficult to run.

Table 6.3　Advantages and disadvantages of cohort and case-control studies

	Advantages	Disadvantages
Cohort studies	• Direct, logical structure • Provides more information • Can control what data is recorded	• Large sample size needed to study rare diseases • Lengthy studies needed when time between exposure and disease occurrence (induction period) long
Case-control studies	• Copes easily with both rare diseases and diseases with long induction periods	• Difficult to find an appropriate control group, i.e. a group that gives a correct description of the exposure frequency in the populations from which the cases are recruited • Measure of exposure difficult since it is performed after the disease has occurred. It is possible that the cases are aware of the hypothesis being investigated and are more likely to report exposure than the controls

Case-control studies are easier to run than cohort studies and require fewer subjects. For this reason, they are often tried first to determine the role of certain risk factors in the development of disease. The optimal proportion of cases to controls is one case to five controls. The main advantages and disadvantages of cohort and case-control studies are listed in Table 6.3.

6.6　Choice of comparison group

The control group in cohort studies has to be as similar to the exposed group as possible with respect to relevant risk factors, except for the exposure under study. In case-control studies we select the control group so that it reflects the presence of exposure in the population from which the cases were selected. The typical approaches to choosing a control group for cohort and case-control studies are given in Table 6.4.

Table 6.4   Choice of comparison group for cohort and case-control studies

	Choice of comparison groups
Cohort studies	1. Internal comparisons: a general cohort containing both exposed and unexposed individuals is followed. 2. External comparisons: an exposed cohort is followed and attempts are made to find a comparison group that does not have the specific exposure, but is otherwise similar to the exposed cohort. 3. Comparisons with the 'general' population: an exposed cohort is followed and compared with the general population.
Case-control studies	1. A sample from a population in a given geographical area provided that the cases represent all cases within that area. 2. If the cases are all recruited from a certain hospital, controls are often chosen from among the patients with other diagnoses in the same hospital

6.7   Confounding

If any cause of the disease, besides the exposure under study, is distributed unequally in the two groups that are being compared, this will in itself give rise to differences in disease frequency in the two groups. This distortion leads to an invalid comparison and is called **confounding**. For example: *There is a strong association between yellow index finger and lung cancer.*

Yellow index finger does not cause lung cancer. Smoking is the cause of both lung cancer and yellow index finger, and is consequently a confounder. Smoking is more common among individuals with yellow index finger and since smoking increases the risk of lung cancer, an increased incidence of lung cancer will be observed in the group with yellow index finger.

The solution to confounding is to use **stratification**. This is introduced in Chapter 2, and for further details on stratification we refer you to Ahlbom and Norell (1984).

6.8 Summary

Our aim in this chapter is to provide a basic introduction to epidemiology. It is our view that epidemiological studies are among the most difficult studies to conduct and interpret, and they should not be attempted without the assistance of a professionally trained epidemiologist.

We include the following list of critical questions to help you interpret epidemiological studies:

1. Size of the relative risk: a large relative risk is less likely to be explained by bias alone.

2. Repeatability: do the results agree with those from other studies?

3. Biological plausibility: is the study linked to a plausible biological mechanism?

4. Epidemiological plausibility: are the results consistent with large national statistics, etc.?

5. Quality of study: tightness of definitions, narrowness of aims, thoroughness, choice of control group.

6. Exposure-disease relationship: for example, are heavier smokers at a greater risk of developing lung cancer than infrequent smokers?

Areas outside the scope of this chapter include *retrospective cohort studies* and *matched case-control studies*. For information on these studies we refer you to Ahlbom and Norell (1984).

7 MEASURING THE AVERAGE

7.1 What is an average?

One meaning of statistics is to do with describing and summarizing quantitative data. Any description of a sample of observations must include an aspect that relates to **central tendency**. That is to say, we need to identify a single number close to the centre of distribution of the observations that represents them all. We call this number the **average**. It is often referred to as a measure of **location** because it indicates, on what might be a scale of infinite magnitude, just where a cluster of observations is located. The average is described by one of three commonly used statistics: the **mean**, the **median**, and the **mode**.

7.2 The mean

The mean or, more precisely, the **arithmetic mean**, is the most familiar and useful measure of the average. It is calculated by dividing the sum of a set of observations by the number of observations. If it is possible to obtain an observation from every single item or sampling unit in a population, then the mean is symbolized by μ (mu) and is called the **population mean**. More usually, we have to be content with the observations from a sample, in which case the mean is symbolized by $\bar{x}$ ('x-bar'). The sample mean is a direct estimate of the population mean (thus, $\bar{x}$ estimates μ). In Chapter 10, we explain how good an estimate it is likely to be. The formulae used for calculating the mean are:

$$\mu = \frac{\Sigma x}{N} \quad \text{and} \quad \bar{x} = \frac{\Sigma x}{n}$$

where x is each observation, N is the number of items (observations) in a population, n is the number of observations in a sample, and Σ means 'the sum of'.

Example 7.1

The survival times (months) of five patients with abdominal cancer are:
8.5   9.2   7.3   6.8   10.1
Calculate the mean survival time.

1. Obtain the sum of the observations (Σx):
   $\Sigma x = 8.5 + 9.2 + 7.3 + 6.8 + 10.1 = 41.9$ months

2. Divide Σx by n, the number of observations:

   $$\bar{x} = \frac{\Sigma x}{n} = \frac{41.9}{5} = 8.38 \text{ months}$$

Conventionally, the mean is recorded to one more decimal place than the original data if n is less than 100, to two decimal places if n is between 100 and 999, and so on.

Example 7.2

The number of casualties received by an accident and emergency department in five one-hour periods is:

28   16   24   31   27

Calculate the mean number of patients per hour:

$$\bar{x} = \frac{\Sigma x}{n} = \frac{126}{5} = 25.2$$

In this example, observations relate to a discrete (non-continuous) variable – a whole number of patients. The sampling unit is a one-hour period. Notice, however, that *the mean is a fraction*; it can assume any degree of precision. Therefore, a sample mean is a *continuous variable*, even if the observations in the sample are not. The consequences of this important point are considered in Section 10.1

7.3 Calculating the mean of grouped data

In Examples 7.1 and 7.2 the mean is calculated from *ungrouped* observations. In cases where a larger number of observations are *grouped* into a frequency table, the calculation of the mean is different. The formula of deriving the mean of grouped data is:

$$\bar{x} = \frac{\Sigma f x}{n}$$

where x is the frequency class, f is the frequency of occurrence in a class and Σ is 'the sum of'. We show how this formula is applied in Example 7.3.

Example 7.3

Calculate the mean of the grouped baby weight data given in Table 3.2. In the table, the first frequency class is 2.5 kg, of which the frequency of occurrence is 1. The first component of the $\Sigma f x$ term is thus (1×2.5). The next frequency class is 2.6, of which the frequency is 2, making the second component (2×2.6). Skipping a few, the sixth frequency class is 3.0, of which the frequency is 15. The corresponding component is therefore (6×15). In this way, all 13 frequency-classes are inserted in the top part of the formula (numerator). In the bottom part of the formula (denominator) is the total number of frequencies, $1 + 2 + 4 + 7$, and so on.

Writing it out in full, the mean is:

$$\bar{x} = \frac{\begin{array}{c}(1 \times 2.5) + (2 \times 2.6) + (4 \times 2.7) + (7 \times 2.8) + (11 \times 2.9) + \\ (15 \times 3.0) + (20 \times 3.1) + (15 \times 3.2) + (11 \times 3.3) + \\ (7 \times 3.4) + (4 \times 3.5) + (2 \times 3.6) + (1 \times 3.7)\end{array}}{1 + 2 + 4 + 7 + 11 + 15 + 20 + 15 + 11 + 7 + 4 + 2 + 1}$$

$$= \frac{310}{100} = 3.1 \text{ kg}$$

7.4 The median – a resistant statistic

The numbers of patients recorded dying of hand-gun injuries in hospitals in various countries in a year per 5 million population are as follows:

Switzerland	Japan	UK	Sweden	Canada	Germany	USA
24	2	1	12	8	3	190

The average number of deaths per country $= \Sigma x/n = 240/7 = 34.3$ per 5 million population. The calculated value of 34.3 scarcely represents all the observations in the sample. It is larger than six of the seven observations, and is more than five times smaller than the other. Because the mean takes into account the value of every observation in a sample, it can be greatly distorted by a single exceptional value. Perhaps hand-gun laws are different in the USA. When a few exceptional values distort the mean in this way, a *resistant measure of the average*, namely the *median*, may be more appropriate.

Example 7.4

Look again at the distribution of the hand-gun counts given above. The observations are ranked in ascending order.

1 2 3   8   12 24 190
          median

The observation 8 has three observations to the left which are smaller and three to the right which are larger. It is the middle value, and the sample *median*. Notice how it is *resistant* to the extreme observation; the median is 8 if the seventh observation 90, 190 or 9000. The median is more *representative* of this set as a whole than the mean. If there is an even number of observations in a sample, there is no middle value. By convention, the median is taken to be the mean of the values of the middle pair.

Example 7.5

Find the median of the following observations:

9.2   11.5   13.2   19.7   29.4   50.1

The median lies between the third and fourth observations. It is estimated by:

$$\text{Median} = \frac{13.2 + 19.7}{2} = 16.45$$

The same value is obtained by halving the *difference* between the two middle observations and adding the result to the lower value, or subtracting it from the larger:

$$\text{Median} = 13.2 + \frac{19.7 - 13.2}{2} = 16.45$$

7.5   The median of a frequency distribution

When there is a large number of observations, it may not be possible to identify a single observation or a 'central pair' with which to define the median. In Example 7.6, the median of the nine observations appears, at first sight, to be '5'. But there are three observations of 5. In this situation, the estimation of the median is a little more complicated.

Example 7.6

Calculate the median of the following observations (mm):

2   3   3   4   5   5   5   6   10

1. Draw a dot plot of the data.

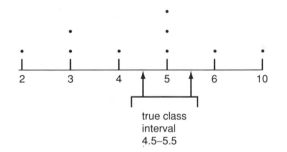

2. The median is $n/2 = 9/2 = 4.5$, that is, the '4.5th' observation from either end of the distribution.

3. Starting from the left, the cumulative frequency in the first three classes is $1 + 2 + 1 = 4$. The median is therefore the '0.5th' observation in the next class. There are three observations in the next class; therefore the median is $0.5/3 = 0.167$th of the way into the class.

4. We recognize that the three observations of 5 mm are not *precise*: they are accurate to within ± 0.05 mm. The observations fall within a class interval with implied lower and upper limits of 4.5 and 5.5 mm.

5. The breadth of the class is $5.5 - 4.5 = 1$ mm. The median is 0.167th of the way into the interval, that is, 0.167 mm. Add this to value of the lower interval limit: $4.5 + 0.167 = 4.667$ mm. This, then, is the median.

6. Confirm the result by working down the distribution from the right. The median observation is the $n/2 = 4.5$th observation. The cumulative frequency down to the median class interval $= 1 + 1 = 2$. The median observation is therefore the 2.5th into the interval. There are three observations in the interval; the median, then, is $2.5/3 = 0.833$th of the way into the interval. Because the interval breadth is 1 mm, 1×0.833 is subtracted from the value of the upper interval limit, $5.5 - 0.833 = 4.667$. The median checks at 4.667 mm.

7.6   The mode

The mode is another measure of the average. In its most common usage this measure is called the *crude mode* or *modal class*. It is the class in a frequency distribution that contains more observations than any other. The modal baby weight class in Section 3.11 is 3.1 kg. The modal class in Example 7.6 is 5. The mode is the only measure of the average that can be used on *ordinal* scales.

Example 7.7

The numbers of cases of heart disease among 100 patients classified according to age class is presented below. What is the average score?

Child	2
Teen	4
Adult	22
Older adult	45
Elderly	14
Older elderly	13

The scale of measurement is related to the age of the patient, and we might be interested in the 'average age' of the incidence of heart disease. However, because the age-scale is ordinal (for example, we cannot say exactly how many times older 'elderly' is than 'teen') we are unable to calculate the mean of the distribution. However, there are more cases in 'older adult' than any other. The modal class is therefore *Older adult* (45), and this is the only measure of the 'average' of this distribution that can be used.

A frequency distribution with more than one peak, or mode, is called a *multi-modal distribution*. When there are two peaks it is a *bimodal* distribution. An example of a bimodal distribution is shown in Figure 7.1. In this figure, although there is overlap the men have a larger average Forced Expiratory Volume and represent the right modal peak.

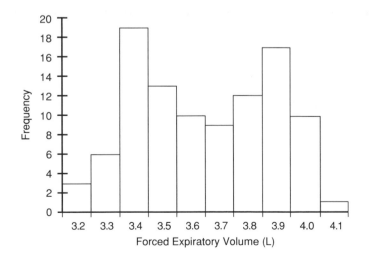

Figure 7.1   Bimodal distribution of Forced Expiratory Volume in one second of 100 healthy 16-year old men and women

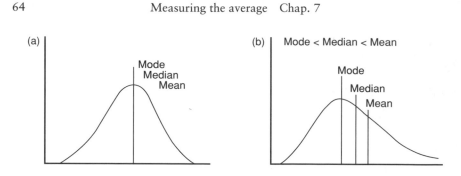

Figure 7.2   (a) Symmetrical distribution; (b) skewed distribution

7.7   Relationship between mean, median and mode

In a perfectly symmetrical distribution the mean, median and mode have the same value. In a skewed distribution, the mean shifts towards the direction of the skew. In biological distributions a skew is nearly always to the right (positive skew), and so the mean is larger than the median or mode, as Figure 7.2 illustrates.

The mean is the only one of the three measures that uses all the information in a set of data. That is, it reflects the value of each observation in the distribution. Its precisely defined mathematical value allows other statistical techniques to be based upon it. Moreover, it has the advantage of being capable of combination with the means of other samples to obtain an overall mean. This may be useful if individuals (sampling units) are rare or hard to come by.

8 MEASURING VARIABILITY

8.1 Variability

If there were no variability within populations there would be no need for statistics: a single item or sampling unit would tell us all that is needed to know about the population as a whole. It follows, therefore, that in presenting information about a sample it is not enough simply to give a measure of the *average*; what is also needed is information about **variability** within a sample. A dot plot is a qualitative method of assessing the variability in a sample, as Figure 4.1 shows. For a quantitative analysis of variability within and between samples, we need a mathematically defined measure. The three that we describe are the **range**, **standard deviation** and **variance**.

8.2 The range

The simplest measure of variability in a sample is the *range*. This indicates the highest and lowest observations in a distribution, and shows how wide the distribution is over which the measurements are spread. For continuous variables, the range is the arithmetic difference between the highest and lowest observations in the sample. In the case of counts or measurements, remember to add 1 to the difference because the range is inclusive of the extreme observations.

The range takes account of only the two most extreme observations. It is therefore limited in its usefulness, because it gives no information about how the observations are distributed. Are they evenly spread, or are they clustered at one end? The range should be used when observations are too few or too scattered to warrant the calculation of a more precise measure of variability, or when all that is required is an appreciation of the overall spread of the observations.

Finding the range is an important step in constructing a frequency distribution (see Section 3.12).

8.3   The standard deviation

The *standard deviation* is the most widely applied measure of variability. It is calculated directly from all the observations of a particular variable. When observations have been obtained from *every* item or sampling unit in a population, the symbol for the standard deviation is σ (lower case sigma). This is a parameter of the population. When it is calculated from a sample it is symbolized by s. In practice, s is easily calculated using a hand calculator, but first we explain how it is done 'long-hand'.

The formula for calculating σ is:

$$\sigma = \sqrt{\frac{\Sigma (x - \mu)^2}{N}}$$

where x = the value of an observation, μ = population mean, Σ = 'the sum of' and N = the number of items in the population.

However, it is rarely possible to obtain observations from every item or sampling unit in a population. For example, to measure the variability in pulse rate in school children, we would not dream of obtaining observations from every child in the country. We would randomly select a suitably sized sample. We have, therefore, to be content with *estimating* σ in the population from the observations in a sample. The symbol for a standard deviation obtained from sample data is s. The formula for obtaining s is:

$$s = \sqrt{\frac{\Sigma (x - \bar{x})^2}{n - 1}}$$

where $\bar{x}$ is the sample mean and n is the number of observations in the sample. The term $(x - \bar{x})$ is known as the **deviation from the mean**. In cases where $\bar{x}$ is larger than x, the deviation is negative; squaring the number always makes it positive. In the formula for s, note that the denominator is $(n - 1)$. The reduction by 1 has the effect of increasing s. This has been described neatly as a 'tax' to be paid for using the sample mean $\bar{x}$ (a statistic) instead of the population mean μ (a parameter) in the estimation of σ. The expression $(n - 1)$ is known as the **degrees of freedom**. We explain the meaning of this expression more fully at the end of the chapter, but in estimating a standard deviation, the degrees of freedom are always one less than the number of observations in a sample. Although s is an estimate of the *population standard deviation*, it is widely referred to as the **sample standard deviation** because it is calculated from sample data.

It is possible to calculate an alternative sample standard deviation by placing n, rather than the degrees of freedom $(n - 1)$, in the denominator. This, however, is a *descriptive statistic only*, and has no meaning in inferential statistics. We recommend that you never use it unless you have a very clear reason for doing so. In practice, you will undoubtedly choose to work out a standard deviation directly from a scientific calculator. This usually has separate keys marked n and $(n - 1)$ for population and sample standard deviations, respectively. In the next section, we show you how to calculate the standard deviation from first principles. It is important that you study this, because it introduces a term called the sum of squares that has application in more advanced statistical methods that we touch upon in later chapters.

8.4 Calculating the standard deviation

The procedure for calculating the standard deviation from first principles is shown in Example 8.1.

Example 8.1

Calculate the standard deviation of the following 10 observations (mm):
81   79   82   83   80   78   80   87   82   82

1. Calculate the mean, $\bar{x}$:

$$\bar{x} = \frac{\Sigma x}{n} = \frac{814}{10} = 81.40 \text{mm}$$

2. Obtain the deviations from the mean by subtracting the mean from each observation in turn; square each deviation (the squaring eliminates any minus signs):

$(81 - 81.4)^2 = 0.16$ $(78 - 81.4)^2 = 11.56$
$(79 - 81.4)^2 = 5.76$ $(80 - 81.4)^2 = 1.96$
$(82 - 81.4)^2 = 0.36$ $(87 - 81.4)^2 = 31.36$
$(83 - 81.4)^2 = 2.56$ $(82 - 81.4)^2 = 0.36$
$(80 - 81.4)^2 = 1.96$ $(82 - 81.4)^2 = 0.36$

3. Add up the 10 squared deviations: the sum of the squared deviations is 56.4. This sum $\Sigma (x - \bar{x})^2$ is called the **sum of the squares of the deviations** or, more simply, **the sum of squares.**

4. Divide the sum of squares by one less than the number of observations:

$$\frac{\text{sum of squares}}{(n-1)} = \frac{\Sigma (x - \bar{x})^2}{(n-1)} = \frac{56.4}{9} = 6.27$$

5. The standard deviation is the square root of this value:

$$s = \sqrt{6.27} = 2.50 \text{mm}$$

We estimate that the sample of 10 observations is drawn from a population whose standard deviation is 2.50 mm.

8.5   Calculating the standard deviation from grouped data

The calculation of the standard deviation of a large sample is less of a chore if the data are grouped. The formula for the standard deviation of grouped data is:

$$s = \sqrt{\frac{\Sigma f (x - \bar{x})^2}{n - 1}}$$

Example 8.2

Calculate the standard deviation of the 100 baby weights of Section 3.11, of which $\bar{x}$ has already been worked out as 3.1 kg:

Table 8.1 Calculating a standard deviation of grouped data

Frequency class x (kg)	$(x - \bar{x})^2$	Frequency f	$f(x - \bar{x})^2$
2.5	0.36	1	0.36
2.6	0.25	2	0.5
2.7	0.16	4	0.64
2.8	0.09	7	0.63
2.9	0.04	11	0.44
3.0	0.01	15	0.15
3.1	0	20	0
3.2	0.01	15	0.15
3.3	0.04	11	0.44
3.4	0.09	7	0.63
3.5	0.16	4	0.64
3.6	0.25	2	0.5
3.7	0.36	1	0.36
		$n = \Sigma f = 100$	$\Sigma f(x - x)^2 = 5.44$

1. Obtain the value of $(x - \bar{x})^2$ for each frequency class, and multiply it by the number of observations in each class. Thus, in the frequency class 2.5, $(x - \bar{x})^2 = (2.5 - 3.1)^2 = 0.36$; $0.36 \times 1 = 0.36$. Sum the columns of f and $f(x - \bar{x})^2$. The results are shown in Table 8.1.

2. $s = \sqrt{\dfrac{\Sigma f(x - \bar{x})^2}{n - 1}} = \sqrt{\dfrac{5.44}{99}} = \sqrt{0.0549} = 0.23$

Scientific calculators have a facility for entering grouped data for computing a standard deviation.

8.6 Variance

An important measure of variability closely related to the standard deviation is *variance*. Variance is used in certain parametric techniques described later. Variance is the square of the standard deviation; conversely, a standard deviation is the square root of the variance. Variance is symbolized σ^2 for a population variance and s^2 for a variance estimated from a standard deviation. Thus:

$$\sigma = \sqrt{\sigma^2} \quad \text{and} \quad s = \sqrt{s^2} \quad \text{and} \quad s^2 = \frac{\Sigma(x - \bar{x})^2}{n - 1}$$

It follows that the variance is the value obtained before taking the square root in the final step of the calculation of the standard deviation. The variance of the sample in Example 8.2 is 0.0529.

8.7 An alternative formula for calculating the variance and standard deviation

The methods described in Sections 8.4 and 8.5 illustrate the principle of calculating the variance and standard deviation. An algebraic rearrangement of the formula is, in practice, easier to handle. Moreover, it introduces two statistical expressions that will appear again later. The alternative formula for obtaining the sum of squares (and hence variance) is:

$$\text{Sum of squares} = \Sigma x^2 - \frac{(\Sigma x)^2}{n}$$

The two new expressions are Σx^2 and $(\Sigma x)^2$

(i) Σx^2 is called *the sum of the squares of x*. Using the 10 measurements of Example 8.1, it is derived as follows:

x: 81   79   82   83   80   78   80   87   82   82

$\Sigma x^2 = 81^2 + 79^2 + 82^2 + 83^2 + 80^2 + 78^2 + 80^2 + 87^2 + 82^2 + 82^2$

$= 66316$

(ii) $(\Sigma x)^2$ is called *the square of the sum of x* and is calculated as follows:

$(\Sigma x)^2 = (81 + 79 + 82 + 83 + 80 + 78 + 80 + 87 + 82 + 82)^2$

$= (814)^2 = 662596$

Substituting in the formula for the sum of squares:

$$\text{Sum of squares} = 66316 - \frac{662596}{10} = 56.4$$

(i) The variance is obtained by dividing the sum of squares by $(n - 1)$:

$$s^2 = \frac{56.4}{(10 - 1)} = 6.27$$

(ii) The standard deviation is the square root of the variance:

$$s = \sqrt{6.27} = 2.50\,\text{mm}$$

8.8 Obtaining the standard deviation and the sum of squares from a calculator

'Scientific' calculators have facilities for entering observations and, by pressing the appropriate keys, obtaining $\bar{x}$, standard deviation, Σx and Σx^2. The calculator has to be set in 'scientific', 'statistical' or 'standard deviation' mode, according to the make of your calculator.

If you are unsure about obtaining these quantities from your calculator, we recommend that you stop at this point and learn to do so from the instruction booklet accompanying the instrument before proceeding further. There is one particular caution that needs to be mentioned here. Most calculators have two keys that will show the standard deviation. According to the make of your calculator, they may be shown as σ_n and σ_{n-1}, or σ and s. These relate to the population and sample standard deviations, respectively. We recommend that you always use the 'sample' standard deviation, unless you are quite sure that you have in fact sampled the entire population.

Calculators do not generally have facilities for obtaining the variance, sum of squares and $(\Sigma x)^2$ directly.

1. To obtain the variance, first obtain the standard deviation, and then square it.

2. To obtain $(\Sigma x)^2$, obtain Σx and square it.

3. To obtain the sum of squares, obtain the variance as in (1) and then multiply by $(n-1)$.

8.9 Degrees of freedom

In obtaining a sample standard deviation to estimate a population standard deviation in Section 8.3, reference is made to the number of *degrees of freedom*. Because the concept of degrees of freedom is involved in many statistical techniques, it now needs a fuller explanation. This simplified description may help.

Suppose that we are told that a sample of $n = 5$ observations has a mean $\bar{x}$ of 50, and we are then asked to 'invent' the values of the observations. We know that $\Sigma x = (\bar{x} \times n) = 250$. If the sum of the observations is 250, we have freedom of choice only for the first four numbers, because the fifth must be a number (perhaps a negative number) that brings the sum to 250. By way of example, if the first four numbers are arbitrarily chosen as, say, 40, 25, 18, 130, then in order to make $\Sigma x = 250$, we have no further freedom of choice; the fifth number must be 37. Degrees of freedom (df) in our present example is one less than the number of observations. That is, $\mathrm{df} = (n - 1)$.

Sometimes the formula for estimating a population parameter contains a value which is itself an estimate. Thus, to estimate a standard deviation, the value of the mean is required. Because the value of the mean is itself is estimated from the sample data, this 'costs' a degree of freedom, which has been compared with a 'tax' to be paid for the approximation.

The degrees of freedom are not always $(n - 1)$; they depend upon the particular estimation in hand. We explain the rules for deciding the degrees of freedom for each technique as it arises. Degrees of freedom are symbolised by the Greek symbol ν (pronounced 'nu').

8.10   The Coefficient of Variation (CV)

The standard deviation s is a measure of the degree of variability in a sample that estimates the corresponding parameter of a population. However, it is of limited value for *comparing* the variability of samples whose means are appreciably different. When comparing variability in samples from populations with different means, the **coefficient of variation (CV)** is used. This is the ratio of the standard deviation to the mean, usually expressed as a percentage by multiplying by 100.

Example 8.3

Aspects of growth in children are investigated. The mean and standard deviation of samples of (a) newborn babies, (b) three-year old children, and (c) ten-year old children are investigated. Does the relative variability change with age?

(a) New born babies: $\bar{x} = 3.1\,\text{kg}$; $s = 0.23\,\text{kg}$.

$$CV = 0.23/3.1 \times 100 = 7.4\%$$

(b) 3-year old children: $\bar{x} = 16.0\,\text{kg}$; $s = 4.5\,\text{kg}$.

$$CV = 4.5/16.0 \times 100 = 28.1\%$$

(c) 10-year old children: $\bar{x} = 35.0\,\text{kg}$; $s = 13.8\,\text{kg}$.

$$CV = 13.8/35.0 \times 100 = 39.4\%$$

It is clear from an inspection of the three values of CV that relative variability increases with age.

9 PROBABILITY AND THE NORMAL CURVE

9.1 The meaning of probability

The mathematical theory of probability arose from the study of games of chance (gambling). That is why so many textbook illustrations of probability involve dice throwing and coin-tossing (though we will try to stimulate your interest with less mundane examples!). Probability may be regarded as quantifying the chance that a stated **outcome** of an **event** will take place. By convention, probability values fall on a scale between 0 (impossibility) and 1 (certainty), but they are sometimes expressed as percentages, so the 'probability' scale has much in common with the 'proportion' scale.

Example 9.1

Among the first 100 donors to present themselves at a blood-donor unit, 34 were of blood group type A. In this example, we define an *event* as 'the result of a blood-group test', and the *outcome* is 'type A' or 'not type A'.

(a) What is the estimated probability that a donor selected at random will be of type A?

$$P = \frac{\text{Number of nominated outcomes}}{\text{Number of possible outcomes}}$$

$$P = \frac{\text{Number of donors tested as Type A}}{\text{Total number of donors tested}} = \frac{34}{100} = 0.34 \ (34\%)$$

(b) What is the estimated probability that a donor selected at random
     will *not* be of type A?

$$P = \frac{\text{Number of donors tested as } not \text{ Type A}}{\text{Total number of donors tested}} = \frac{66}{100} = 0.66 \ (66\%)$$

Notice first that the estimated probabilities are equal to the *proportional
frequency* of an item in the sample (type A or *not* type A, respectively –
Section 3.7). Secondly, the sum of the separate probabilities
(0.34 + 0.66) equals 1. It is therefore certain that a donor will be, or
will not be, of blood group type A.

9.2 Compound probabilities

Probability values may be added and multiplied. Because probability
values are fractions, adding probabilities *increases* the likelihood of a
stated outcome whilst multiplying probabilities *decreases* it. The decision whether to add or multiply may therefore often be made intuitively.

Example 9.2

We extend the scenario of Example 9.1 by supposing that the donors
were additionally classified into age groups, namely over 25 years old
(the *older* group) under 25 (the *younger* group). The composition of the
sample is as follows:

	Older		Younger		Total
	Type A	Not type A	Type A	Not type A	
	20	40	14	26	
Totals	60		40		100

The probability of a nominated outcome (e.g. *older, younger, older
type A,* etc.) can be worked out by a simple tree diagram in which
the branches show the proportional frequencies (probabilities) of a
particular outcome. Thus, the probability of *older* is 60/100 = 0.6; the

probability of *younger* is 40/100 = 0.4. Of the *older* group, 20/60 (0.33) are of *type A*, while 40/60 (0.67) are *not* of *type A*.

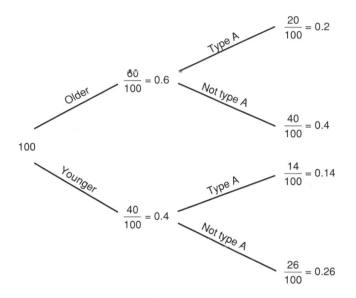

What is the probability that a donor selected at random will represent *either* the '*older*' or the '*younger*' group? Clearly, the probability of this outcome is greater than one, in which only a single group is nominated:

Probability of outcome *younger*   = 0.4
Probability of group *older*   = 0.6
Probability of outcome *older* or *type A*   = 0.4 + 0.6 = 1.0

What is the probability that a donor selected at random will be *older* and *not type A*? In this case, since *two* conditions (age group and blood type) are stipulated, the probability is lower than for any single condition:

Probability of outcome *older*:   = 0.6
Probability of outcome *not type A*   = (0.4 + 0.26) = 0.66
Probability of outcome *older* and *not type A*   = 0.6 × 0.66 = 0.4

Example 9.3

If the outcome of an event is described in terms of the age group and blood group type of a single donor selected at random, show that the sum of the probabilities for all possible outcomes is equal to 1.

	Probability of outcome		
Older type A	Younger type A	Older not type A	Younger not type A
$0.6 \times (20/60) =$ 0.2	$0.4 \times (14/40) =$ 0.14	$0.6 \times (40/60) =$ 0.4	$0.4 \times (26/40)$ 0.26

Total probability $= 0.2 + 0.14 + 0.4 + 0.26 = 1$

A breakdown which shows the individual probabilities of all possible outcomes and which adds up to 1 is called a **probability distribution**.

9.3   Critical probability

One of the important considerations in statistics is to know the probability below which a stated outcome is considered unlikely to be the result of chance alone. Such a value is called a **critical probability**, and is decided arbitrarily by statisticians.

Consider the following imaginary scenario. A blood transfusion department is keen to improve stocks of a particular blood group type. The type is known to occur, on average, in 1 in 10 members of the community. In other words, the probability that a donor selected at random is of that blood group type is $1/10 = 0.1$. How should a blood donation unit react if:

(a) The first donor was of the required type?

(b) the first *two* donors were of the required type?

(c) the first *three* donors were of the required type?

(a) If the first donor proved to be of this type, the unit should probably feel a little lucky, but there would be no cause for major surprise. This outcome is, after all, to be expected from one trial in 10.

(b) The probability that the first two donors are of the required type is $0.1 \times 0.1 = 0.01$, which is to be expected from only one trial in 100. The unit should feel most fortunate that a further specimen is obtained so easily. They might even harbour a suspicion that the blood group types in the community are not truly independent (e.g. a family had arrived together to donate blood).

(c) The probability that the first three donors are of the required blood group type is $0.1 \times 0.1 \times 0.1 = 0.001$, i.e. the predicted outcome from one trial in a thousand. If this actually happens, the unit has very strong grounds indeed for considering that the selection of the donors is not truly random. (Perhaps the word had leaked out that the particular blood group was required, and donors of the required type had presented themselves in greater numbers.)

At what point does the probability of an outcome become so low that it must be regarded as 'unlikely'? Statisticians conventionally adopt three critical probability values:

1. An outcome that is predicted to occur in less than 1 trial in 20 ($P < 0.05$) is considered to be *unlikely* or *statistically significant**.

2. An outcome that is predicted to occur in less than 1 trial in 100 ($P < 0.01$) is considered to be *very unlikely* or *statistically highly significant***.

3. An outcome that is predicted to occur in less than 1 trial in 1000 ($P < 0.001$) is considered to be *extremely unlikely* or *statistically very highly significant****.

The three levels are often indicated *, ** and ***, respectively.

9.4 Probability distribution

Section 9.2 concluded with an example of a simple probability distribution. Probability distributions may be generated empirically, that is, by sampling and observation. Thus, recalling Figure 4.6 that is a frequency distribution of 100 new-born baby weights, we can convert a frequency distribution into a probability distribution by re-scaling the vertical axis and dividing by the number of observations (sample size). This has been done in Figure 9.1. It may be seen that, for example, the probability that a baby selected at random has a weight that falls in the 3.1 kg class is 20/100 = 0.2. Moreover, if the individual probabilities for each frequency class are summed, the result is 1.

Why are probability distributions useful? First, as indicated above, they allow us to estimate the probability that an event (e.g. the random selection of a baby in this example) will have a stated outcome (e.g. a

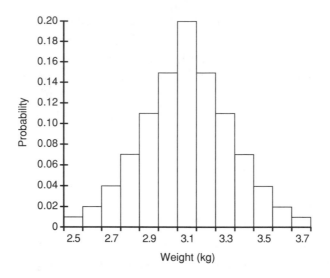

Figure 9.1   Estimated probability distribution of new-born baby weights

weight falling within the 3.1 kg class). Secondly, a probability distribution may be used to generate a distribution of **expected frequencies**. Just as a probability is estimated by dividing a frequency of a particular observation by the total number of observations, multiplying a probability by the number of observations produces an expected frequency:

Expected frequency = (estimated probability) × (number of observations)

Example 9.4

Using the probability distribution worked out in Example 9.3, what is the *expected frequency* of each outcome in terms of age group and blood group type in a sample of 240 donors in another community (rounded to whole numbers)?

Expected frequency of *Older type A* $= 0.2 \times 240 \ = \ 48$
Expected frequency of *Younger type A* $= 0.14 \times 240 = \ 34$
Expected frequency of *Older not type A* $= 0.4 \times 240 \ = \ 96$
Expected frequency of *Younger not type A* $= 0.26 \times 240 = \ 62$

Total expected frequency $= 240$

In practice, the obvious thing to do next is to compare the expected frequencies of each outcome with actual or *observed* frequencies. If the observed frequencies disagree with those expected, there could be a basis for exploring factors which underlie the discrepancy.

9.5   The normal curve

In Figure 4.6 we show the distribution of 100 new-born baby weights measured to a precision of ± 0.05 kg. Because measurements of weight are on a continuous scale, they may be made with increasing precision to, say, ± 0.005 kg. If the number of observations is accordingly increased, the steps in the histogram are more gradual, as shown in Figure 9.2(a). As increasing numbers of observations are obtained with increasing degrees of precision, the histogram verges towards a smooth, symmetrical bell-shaped curve, as shown in Figure 9.2(b).

The mathematician Gauss discovered that distributions of this shape estimated from large numbers of measurements could be described by a complicated looking formula that need not concern us here. Suffice to say that the shape of the curve is governed by two parameters that we have already described, namely the mean (μ) and the standard deviation (σ). Usually, we do not know the values of these population parameters, and so have to estimate the corresponding statistics ($\bar{x}$ and s, respectively) from sample data.

The important point is that *all* normal curves are defined in terms of these parameters, and altering the values just changes the precise shape of each curve (Figure 9.3). Because so much of the data we collect will fit the shape of the normal curve, its underlying properties are of great interest, and so long as we are able to calculate the mean and standard

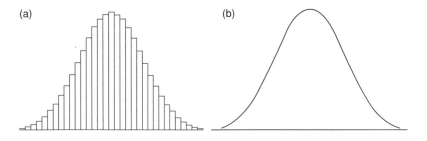

Figure 9.2   Gradation of histogram (a) into the normal curve (b)

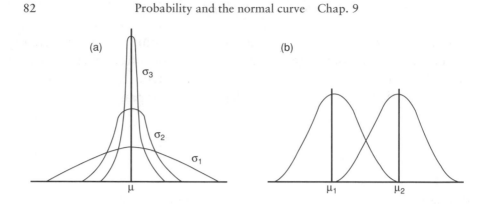

Figure 9.3   Normal curves. (a) Similar μ and different σ; (b) different μ and similar σ

deviation of a sample, it allows us to make all sorts of inferences about the population from which the sample is drawn.

9.6   Some properties of the normal curve

A normal curve is symmetrical, with the axis of symmetry passing through the baseline at one of the parameters – the mean of the distribution. Theoretically, the two tails of the curve never actually touch the horizontal axis (the baseline); rather, they tend to approach it over an infinite distance. In real life, however, the tails are never very long – we do not come across members of the same population that are both astronomically large and microscopically small.

On each side of the curve there is a point of inflexion where the direction of the curve changes from concave to convex. It so happens that if a vertical line is dropped down from the point of inflection, it cuts the baseline at a distance on either side of the central axis equal to the second parameter – the standard deviation. This distance can be used as a standard unit to divide up the baseline (the x-axis) into equal segments, as shown in Figure 9.4.

If the vertical axis of the distribution is rescaled by dividing by the number of observations, it effectively becomes a probability distribution or, strictly, a **probability density**. The total probability encompassed by the density is 1 (100 percent). If we say that the total *area* under the curve is 100 percent, then one of the properties of the normal curve is that the area bounded by one standard deviation on each side of the central axis of symmetry is approximately 68.26 percent of the total area.

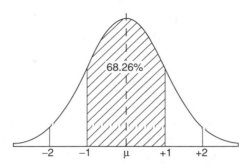

Figure 9.4   Normal curve with standard deviations

This means that about 68 percent of observations drawn from a normally distributed population fall between ± 1 standard deviation from the mean. The other 32 percent fall outside these limits, 16 percent above one standard deviation and 16 percent below one standard deviation. In other words, there is a probability of about 0.68 that a single observation drawn at random will fall between μ ± σ. By extending the limits to *two* standard deviations (μ ± 2σ), the proportion of observations that will be included within them increases to 95.44 percent. Similarly, 99.74 percent of observations fall within μ ± 3σ. These are equivalent to probabilities of 0.9544 and 0.997. In practice, the values of 0.95 and 0.99 are more convenient to deal with. It can be calculated that these probability values fall at μ ± 1.96σ and μ ± 2.58σ, respectively. It will be recalled from Section 9.3 that these are the critical probability values for assessing whether the outcome of a stated event is *unlikely* or *very unlikely*. Therefore, we may use the properties of the normal curve to assess the likelihood of certain outcomes, as we show in the next section.

9.7   Standardizing the normal curve

Any value of an observation x on the baseline of a normal curve can be standardized as a number of **standard deviation units** the observation is away from the population mean μ. This value is called a *z*-**score**. To transform x to z apply the formula:

$$z = \frac{(x - \mu)}{\sigma}$$

If μ is larger than x, then z is negative.

It is often the case that we do not know the values of μ and σ. In samples of more than about 30 observations, z is given by:

$$z = \frac{(x - \bar{x})}{s}$$

By **standardizing** an observation x into a z-score, we can relate it to the properties which apply to all normal curves. Thus, if the calculated value of z is larger than 1.96, then the probability of such a number being drawn at random is less than 0.05. It is regarded as *unlikely* or **statistically significant.**

Example 9.5

On the basis of very large samples of records, the mean weight (μ) of a 'population' of normal healthy new-born babies is estimated to be 3.8 kg and the standard deviation (σ) is estimated to be 0.5 kg. Is it likely that a randomly selected baby of weight 2.0 kg belongs to this population?

Convert the observation x to a z-score:

$$z = \frac{(x - \mu)}{\sigma} = \frac{(2 - 3.8)}{0.5} = -3.6$$

Ignoring the negative sign, the calculated value of z is much larger than 1.96, which is equivalent to $P = 0.05$. Indeed, it is larger than 2.58, larger than 2.58, which is equivalent to $P = 0.01$. We say that the observation is **statistically highly significant,** and conclude that it is very unlikely that the baby belongs to the same population. Babies of this weight clearly fall below that considered typical, and additional post-natal care might be considered appropriate.

This computation and our resulting conclusion is an example of a simple statistical test. We will recall it when we outline the principles of statistical testing in Chapter 11.

9.8   Two-tailed or one-tailed?

In Section 9.6 we said that under a normal curve $\mu \pm 1.96\sigma$ contains 95 percent of observations, and that the residual 5 percent is divided equally outside these limits in both tails of the distribution. This is shown in Figure 9.5.

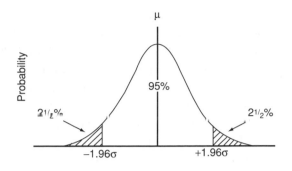

Figure 9.5   The normal curve with 5 percent in two tails

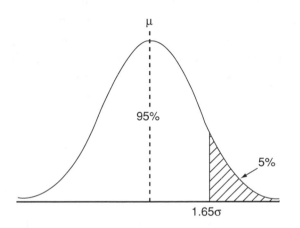

Figure 9.6   The normal curve with 5 percent in upper tail

There is an alternative way of apportioning 95 percent of the observations, and that is by excluding the residual 5 percent in a single tail (either the *upper* tail or the *lower* tail). In this case, the cut-off point occurs at 1.65 standard deviations on whichever side of the distribution the 5 percent is excluded. This is shown in Figure 9.6.

Sometimes we are able to say that if a random observation probably does *not* belong to a population of specified mean and standard deviation, then it *must* belong to a population that has a mean which is specified as being larger or, alternatively, to a population that has a mean which is specified as being smaller. In this case, the critical value of z is set at the lower value of 1.65, and the test is said to be *one-tailed*. Reference to Example 9.5 may help to make this clear. Before the item in question is measured, the question is posed: 'Is its size consistent with a population of specified mean and standard deviation?' If the test shows that it is *not* consistent, then there is an equal chance that it could come from a population whose mean is either larger or smaller. The test is therefore *two-tailed* and the critical value of *z* is set at 1.96.

One-tailed tests may be encountered in screening procedures. Suppose, in Example 9.5, that *only* babies who were *below* a certain critical weight threshold (say, the lowest 5 percent of the distribution) are required to be identified. In this case, the critical threshold could be set at 1.65 standard deviations below the mean (in this example, that works out to be 2.98 kg). This implies that all individuals that are significantly *larger* than the mean are of no interest in the procedure, and are ignored.

Note that the critical value of *z* in a one-tailed test (1.65) is considerably lower than that in a two-tailed test (1.96). Often it is not possible to be sure that a test is genuinely one-tailed. We are usually interested in deviations that are significantly larger *and* smaller than the mean. If you are not sure, always assume that a test is two-tailed and set the critical value of *z* at 1.96 (or higher, e.g. 2.58 if greater stringency is required).

9.9   Small samples: the *t*-distribution

In the normal distribution the *z*-scores of 1.96 and 2.58 indicate the limits on either side of a population mean within which 95 percent and 99 percent of all observations will fall. Usually we do not know the value of μ and σ and are obliged to estimate them from a sample. We can only be confident that a sample mean and standard deviation are reliable estimates of the population parameters if the sample is large. When the sample is small, we are less confident. To compensate for the uncertainty the values of 1.96 and 2.58 are increased, that is, set further out from the mean and the symbol *z* is replaced by the symbol *t*. As the sample size decreases, so too does the degree of certainty. The values of *t* must therefore increase accordingly. Thus, *t*-distributions are dependent on sample size. However, it is not *n* that determines *t* directly, but $(n - 1)$, that is, the degrees of freedom. There are mathematical techniques available for working out the complete probability distributions of *t* for any number of degrees of freedom. In practice we usually need to know only the values which correspond to $P = 0.05$ and $P = 0.01$ for a particular sample size, that is, the values which correspond to the *z*-scores of 1.96 and 2.58 in large samples. These values are found in statistical tables (Appendix 2 in this book).

In the same way that an observation can be converted into a z-score if the estimates of μ and σ are available from a large sample, so too, an observation x may be transformed into a t-score when the values of the mean and standard deviation of a small sample are known. Thus:

$$t = \frac{(x - \bar{x})}{s}$$

If the calculated value of t is larger than that tabulated for $(n - 1)$ degrees of freedom at $P = 0.05$, it is concluded that the observation is unlikely to have been drawn from a population with the same mean as that from which the sample is drawn. In the examples that follow, we refer to the tables for t in Appendix 2.

Example 9.6

How many standard deviations on either side of a sample mean derived from 10 observations would be expected to contain (i) 95 percent and (ii) 99 percent of the observations of the normally distributed population from which the sample is obtained.

(i) If the sample comprises 10 observations, there are 9 degrees of freedom. Enter the table in Appendix 2 at 9 df and run along until the column headed 0.05 (two-tailed test) is reached. The value is 2.262. Therefore, 95 percent of the observations in the population fall between 2.262 standard deviations on either side of the sample mean, that is, $\bar{x} \pm 2.262s$.

(ii) Similarly, the value of t for 99 percent is found in the column headed 0.01 against 9 df, that is 3.250.

Example 9.7

A patient's temperature was recorded hourly from 06.00 hrs to 18.00 hrs. The observations (°C) recorded are:

36.8 37.2 37.9 38.1 38.2 38.1 38.2 37.9 37.6 37.4 37.1 36.9

(Note that these 12 observations comprise a sample drawn from a hypothetical population of all the temperature measurements that *could* have been made between those times.)

The next day the patient's temperature is recorded and found to be 36.0°. Is it likely that the patient's temperature is in the same 'state' as the previous day? (In other words, we are asking if the single observation of 36.0 is likely to belong to the same 'population' of temperatures from which the previous day's sample of 12 was taken.)

Using a calculator, we determine that the mean and standard deviation of the sample of 12 temperatures are 37.62 and 0.52, respectively. Convert the single observation from the next day into a *t*-score:

$$t = \frac{(x - \bar{x})}{s} = \frac{(36.0 - 37.62)}{0.52} = -3.12$$

Because we are using two-tailed tables we ignore the minus sign. Enter the table in Appendix 2 at $(12 - 1) = 11$ df. Under the heading 0.05 we find the value 2.201; our calculated value of 3.12 exceeds this. We conclude that it is unlikely that the single temperature belongs to the 'population' sampled during the preceding day. We can say that it is 'statistically significantly different'. We might infer that the patient's temperature is returning to normal.

Examination of the *t*-tables for $P = 0.05$ shows that as the degrees of freedom increase, the value of t decreases until it verges towards the value of z (1.96) when $n = \infty$ (infinity). Above 30 df, however, the relative change in the value of t becomes very small with increasing sample size; it is very slightly larger than 2.0. The value of z at $P = 0.05$ is only slightly smaller than 2.0. Thus, when the sample size exceeds about 30, the difference between z and t may often be ignored. In fact, the arithmetic convenience of $\bar{x} \pm 2$ standard deviations is often used as the critical limit unless the samples are very small.

9.10   Are our data 'normal'?

Many parametric statistical techniques that we describe depend upon the mathematical properties of the normal curve. They usually assume that samples are drawn from populations that are normally distributed. Sometimes, if samples are small, it is hard to know if the parent population is normal, in which case **distribution free (non-parametric)** techniques are appropriate. Often a simple check will reveal a serious departure from normality. There are, however, some types of samples that can never be regarded as having been drawn from normally-distributed populations.

Because the normal curve is continuous, samples that comprise units of *count* data can never, in theory, be normal. However, the error due to the lack of continuity (that is, due to the 'steps' in a frequency distribution) need not be serious if the distribution is fairly symmetrical. Then we may regard the distribution as being *approximately normal*. Errors are more likely to be serious when the data are strongly skewed.

Even samples of 'measurement' data may depart seriously from normality under some circumstances. For examples, distributions of height measurements of teenagers may have two peaks, corresponding to males and females. Before any sensible statistical analysis can be carried out, the individual classes should be separated out, and analyses conducted on each class separately.

How can we check if samples are likely to have been drawn from a normally distributed population? There are sophisticated statistical methods available to do this, but in practice there is a simpler alternative: plot out the data and see if they *look* normal! As a back up, calculate the mean and standard deviation of the sample and see if about 70 percent of the observations fall within the interval $\bar{x} \pm s$.

Example 9.8

Three different samples with 10 observations in each are shown below. Do they appear to have been drawn from normally distributed populations?

(a)			(b)			(c)	
x	f		x	f		x	f
6	1		5	3		5	3
7	1		6	3		6	2
8	3		7	2		7	1
9	2		8	1		8	2
10	2		9	1		9	2
11	1						

(a) *A dot-plot shows:*

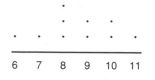

$\bar{x} = 8.6; s = 1.5$

$\bar{x} \pm s$ is about 7.1 to 10.1 which contains 7/10 observations.

   Although the sample is slightly skewed, the observations are scattered on either side of a mode, and 7 out of 10 observations fall within $\bar{x} \pm s$. There is no reason to doubt that the sample is drawn from a normal population.

(b) *A dot-plot shows:*

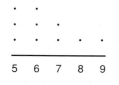

$\bar{x} = 6.4; s = 1.35$

$\bar{x} \pm s$ is about 5.05 to 7.75 which contains 5/10 observations.

   In this case there is a perceptible skew in the data. Only half the observations fall within $\bar{x} \pm s$. We doubt that the sample is drawn from a normally distributed population.

(c) *A dot-plot shows:*

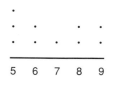

$\bar{x} = 6.8; s = 1.62$

$\bar{x} \pm s$ is about 5.18 to 8.42 which contains 5/10 observations.

   In this case, there is marked bimodalism. $\bar{x} \pm s$ contains only half the observations. We doubt that the sample is drawn from a normally distributed population.

9.11   Dealing with 'non-normal' data

If checks such as those suggested in the previous section show that your data are unlikely to be normally distributed, the question arises as to how they should be correctly analysed. There are two possible solutions. The first is to use the 'distribution-free' **non-parametric** techniques that we describe in later chapters. In most of these, the observations are first converted into *ranks*, thus making the need for a normal distribution unnecessary.

Secondly, we may *normalize* the data by a process of **data transformation**. This process is not as alarming as it might at first sound. It simply means converting each observation into another number by a simple process. The most common transformation is the **logarithmic transformation** (for an introduction to logarithms, see Section 3.8). Thus, each number is replaced by its logarithm, and any subsequent analysis is performed on the logarithms, rather than the original numbers. As an example of how a logarithmic transformation can 'normalize' a set of data, look at Figure 9.7; Figure 9.7(a) shows a frequency bar graph of some count data. Thus, 4 sampling units contain 4 items being counted, 6 contain 5, 8 contain 6, and so on. Notice how the left-hand edge of the horizontal scale is truncated at zero. It is impossible to have a sampling unit with fewer than zero items being counted, so there is a natural absolute end of the scale. On the other hand, it is possible to have (in theory) any number of objects in a sampling unit, so the right-hand end of the scale can stretch away far to the right, giving rise to a **positive skew** to the distribution. The shape of the distribution appears very different from the symmetrical shape we expect of a 'normal' distribution.

In Figure 9.7(b) we have replaced the horizontal scale by the *logarithms* of the observations; that is to say, we have created a *logarithmic scale*. Notice how the right-hand 'tail' of the distribution has been squashed up, making the overall shape more symmetrical – much more like the shape we expect to see in a normal distribution. We say that the data have been *normalized*.

If there are **zero observations** (not uncommon with count data), it is not possible to get a log of zero (try it and see!). To get round this problem, add 1 to all observations. So zero becomes 1; 1 becomes 2; 4.26 becomes 5.26, and so on. Then continue with the analysis as before.

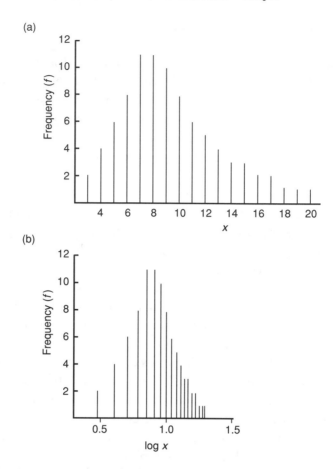

Figure 9.7 (a) Untransformed observations; (b) logarithmically transformed observations

We note at the end of Section 3.7 that *proportions* may have to be transformed before further analysis. That is because the proportional scale is truncated at *both* ends of the scale, at zero and at 1. If the mean proportion in a sample is close to 0.5, then the distribution may be fairly symmetrical; however, the closer it approaches to 0.1 or 0.9, the more skewed the distribution becomes. A suitable transformation for proportional data is the **arcsine transformation**. This is a little more complex than the logarithmic transformation, because two steps are involved. First, make sure your calculator is set in *degrees* mode. Then use it to obtain the square root of x. Secondly, find the angle whose sine equals this value. On most calculators it is found by pressing the $\sin^{-1}$ (inverse sine) button.

Thus, if $x = 0.25$,
$$\sqrt{x} = 0.5$$
$$\sin^{-1} = 30°$$
Therefore, arcsine $0.25 = 30°$

Notice that the scale of the transformation is *angular degrees*. This may seem an awkward scale, but parametric techniques may now be safely performed on proportions that have been arcsine-transformed. If your observations are percentages, divide each by 100 before transforming.

10 HOW GOOD ARE OUR ESTIMATES?

10.1 Sampling error

Usually we obtain a sample in order to derive a statistic (a mean, for example) from the observations that constitute the sample. The sample statistic gives an estimate of the corresponding parameter; thus $\bar{x}$ estimates μ.

Intuitively, we expect that large samples provide more reliable estimates of parameters than small samples and, conversely, that small samples are less reliable than large ones. Several small samples drawn from the *same* population generally provide different values of the same statistic, yet they are all estimates of the same population parameter. The variation between these individual estimates is due to **sampling error**.

Sampling error arises because some samples have, by chance, more than a 'fair share' of larger units whilst others may have more than a 'fair share' of smaller units. Sampling error is not a mistake or error due to an observer; rather it reflects the random scatter inherent in any sample. In a collection of small samples drawn from the same population, some values of the statistic underestimate the parameter, whilst others overestimate it. The way in which the sample statistics cluster around a population parameter is called the **distribution of the statistic** or, sometimes, the **sampling distribution**. Distributions of sample statistics conform to mathematical principles that allow us to state the confidence we may place in estimates of particular population parameters.

10.2 The distribution of a sample mean

The 100 observations of baby weights in Section 3.11 are presented again in Table 10.1, together with the mean of each row. The grand mean of the whole sample has already been worked out as 3.10 kg.

Table 10.1   Weights of 100 babies (kg)

Baby weight (kg)										Mean $\bar{x}$
3.4	3.2	3.0	2.8	2.8	3.1	3.1	3.1	3.1	3.1	3.08
3.3	2.9	3.4	2.5	3.0	3.5	3.1	3.0	3.2	3.6	3.15
2.8	3.1	3.3	3.1	3.4	2.9	3.2	2.8	3.3	3.2	3.11
3.1	2.9	2.6	3.0	3.1	3.1	3.0	3.0	2.9	2.9	2.96
2.9	2.8	3.1	3.3	3.0	2.8	3.1	3.5	3.2	3.3	3.09
3.4	3.0	3.3	3.1	2.7	3.0	3.2	3.6	2.9	2.9	3.11
3.5	3.3	3.1	3.1	3.2	2.9	3.0	3.4	2.7	3.2	3.14
3.0	3.4	3.2	3.0	2.6	3.3	3.5	3.1	2.8	3.4	3.13
3.3	2.7	3.0	3.2	3.2	2.9	3.1	2.7	3.1	3.0	3.02
3.2	3.3	3.0	2.9	3.2	3.3	3.2	3.7	3.2	3.1	3.21

Scrutiny of the 10 row means shows that they vary, ranging from 2.96 kg to 3.21 kg, with not a single one equal in value to the grand mean.

We can group the row-means into a frequency distribution, just as we grouped the observations themselves into a frequency distribution in Table 3.2 (Section 3.11):

Frequency class	Tallies	Frequency f
2.95–3.04	11	2
3.05–3.14	111111	6
3.15–3.24	11	2

Although there are only 10 of these sub-set means, a clear, symmetrical pattern is emerging. Had we more sub-set means drawn from a larger sample, their distribution would resemble that shown in Figure 9.2(a). Because the sample mean is a continuous variable, the distribution verges towards a smooth continuous curve, that is, the normal curve of distribution. This conclusion follows from a fundamental principle in statistics:

> **The Central Limit Theorem** states that the means of a large number of samples drawn randomly from the same population are normally distributed and the 'mean of the means' is the mean of the population.

This normal distribution has it its own standard deviation, that is, a **standard deviation of sample means**. The standard deviation of a set of sample means is given its own name: the **standard error of the mean**.

It is important to note that the Central Limit Theorem makes no assumptions about the underlying population from which samples are drawn. That distribution does *not* have to be normal; it may be symmetrical, skewed or bimodal; observations may be continuous or discrete. Whichever is the case, the means of a large number of samples are approximately normally distributed around the population mean μ.

The properties of the normal curve described in Chapter 9 hold true for a normal distribution of sample means. Thus, about 68 percent of a large number of sample means fall within ± standard error (S.E.) of the population mean μ. The converse of this is also true: we are similarly confident (68 percent) that a *population* mean falls within ± S.E. estimated from a *sample* mean $\bar{x}$.

In practice, we do not estimate a standard error by looking at the spread of sub-sets of sample means. Instead, it may be calculated from the observations of a sample by:

$$\text{S.E.} = \frac{\text{sample standard deviation}}{\sqrt{\text{number of observations}}} = \frac{s}{\sqrt{n}}$$

Example 10.1

Calculate the standard error of the mean of the 100 new-born baby weights for which we have already worked out the mean (3.10 kg) and standard deviation (0.23 kg) in Section 8.5.

$$\text{S.E.} = \frac{s}{\sqrt{n}} = \frac{0.23}{\sqrt{100}} = \frac{0.23}{10} = 0.023$$

In plain language this means: the mean weight of the sample of babies is 3.10 kg and the standard error of the mean is ± 0.023 kg. We are therefore 68 percent confident that the mean of the *population* lies between 3.10 + 0.023 kg (= 3.123 kg) and 3.10 − 0.023 kg (= 3.077 kg). This is, of course, more informative than the standard deviation (s) because it indicates how close the sample mean is likely to be to the population mean which is what we seek to estimate.

10.3   The confidence interval of a mean of a large sample

The standard error of the mean gives us an indication of how good an estimate the sample mean, $\bar{x}$, is of a population mean, μ. Thus, we are 68 percent confident that μ lies within ± 1 S.E. of $\bar{x}$. However, 68 percent is a rather low level of confidence; we usually want to be surer that a population mean lies between indicated limits. To meet this need, 95 percent or 99 percent limits are generally used. These can be obtained by multiplying the appropriate z-score as follows:

- we are 95 percent confident that a population mean falls within $\pm$ 1.96 S.E. of a sample mean;

- we are 99 percent confident that a population mean falls within $\pm$ 2.58 S.E. of a sample mean.

1.96 and 2.58 are the same z-scores that are used in describing the properties of the normal curve (Section 9.6), and are valid only in samples containing about 30 or more observations. The intervals $\bar{x} \pm 1.96$ S.E. and $\bar{x} \pm 2.58$ S.E. are called **95 percent** and **99 percent confidence intervals**, respectively. The adjustments that are made in the case of small samples are described in the next section.

Confidence intervals can be used in two ways. First, they provide a *likely interval* for the true population mean, that is, the 'best guess', given the sample data, of an interval in which the true population mean can be found. Secondly, the length of the interval provides information on the precision that can be attached to the sample estimate. The wider the confidence interval, the less precise is the estimate, and vice versa.

Note that the confidence interval should *not* be interpreted as giving the probability that the population mean falls within a given interval. For example, it is wrong to suggest that the probability of a population mean falling within a 95 percent confidence interval is 0.95. The population mean is an exact number which either does, or does not, fall within the confidence interval.

Example 10.2

Estimate the 95 percent confidence interval of the mean of the 100 baby weights given in Section 3.11.

We have determined in Example 7.4 that the mean of the sample is 3.1 kg and the standard error is ± 0.023 kg. The 95 percent confidence interval is therefore $3.1 \pm (1.96 \times 0.023) = 3.1 \pm 0.045$ kg. This means that we are 95 percent confident that the population mean lies between 3.145 kg and 3.055 kg.

Notice that because n, the number of observations, is the denominator (bottom part) of the equation for estimating the standard error (and hence the confidence interval), the value of the standard error (and the breadth of the confidence interval) gets smaller as n gets larger. This is a mathematical expression of the statistical axiom that the larger the sample size, the greater is the reliability of an estimate of a population parameter.

10.4   The confidence interval of a mean of a small sample

In calculating the standard error of a mean (and hence a confidence interval), the standard deviation is used. Strictly, this should be the *population* standard deviation, σ. In large samples we can be confident that the sample standard deviation, s, is a reliable estimate of σ. We are less certain, however, in the case of small samples. Therefore, we need to apply a correction factor to compensate for the uncertainty in small samples. That factor should become larger as the sample becomes smaller. A suitable correction factor is t, as described in Section 9.9.

Example 10.3

Calculate the mean and 95 percent confidence interval of the following sample of observations of the weights of children (kg):

19.4   21.4   22.3   22.1   20.1   23.8   24.6   19.9   21.5   19.1

Using a scientific calculator, it is determined that $\bar{x} = 21.42$ and $s = 1.84$. The standard error is:

$$S.E. = s/\sqrt{n} = 1.84/\sqrt{10} = 0.582$$

In a small sample we do not use the z-score of 1.96 to obtain the confidence interval, but the value of t found in tables at the appropriate number of degrees of freedom $(n - 1)$. Thus:

$$95\% \ confidence \ interval = \bar{x} \pm (t \times S.E.)$$

In Appendix 2, we find that for $(10 - 1)$ d.f. at $P = 0.05$ (95%), $t = 2.262$. The 95 percent confidence interval is therefore:

$21.42 \pm (2.262 \times 0.582)$, i.e. 21.42 ± 1.32 kg.

Thus, we are 95 percent confident that the mean mass of the population from which the sample is drawn lies between 22.74 kg (upper limit) and 20.1 kg (lower limit). If we had used the z-score 1.96 instead of the t-score of 2.262, the limits would have been 22.56 kg and 20.28 kg – an important reduction in range of about a third of a kg.

10.5   The difference between the means of two large samples

We often wish to compare the means of some variable in two samples. To know how much warmer, faster, lower, heavier or better one sample is from another allows us to make inferences about the populations from which the samples are drawn.

Example 10.4

Systolic blood pressure measurements of samples of patients with and without kidney disease were recorded (30 in each sample). The results are tabulated below:

Without kidney disease: $\bar{x} = 130.2$ mmHg; $s = 19.4$
With kidney disease: $\bar{x} = 149.2$ mmHg; $s = 21.2$

Employing the technique described in Section 10.4, we estimate the 95 percent confidence interval (C.I.) for each mean, using the z-score of 1.96 for a sample size of 30 ($= 29$ d.f.):

Without kidney disease: 95% C.I. $= 130.2 \pm \left(1.96 \times \dfrac{19.4}{\sqrt{30}} \right) =$

$$130.2 \pm 6.94$$

With kidney disease: 95% C.I. $= 149.2 \pm \left(1.96 \times \dfrac{21.2}{\sqrt{30}} \right) =$

$$149.2 \pm 7.59$$

These results are expressed graphically in Figure 10.1. The difference between the two sample means is $(149.2 - 130.2) = 19.0$ mmHg. But what are we to infer about the difference in size between the two populations from which the samples are drawn?

The problem, of course, is that we only have *estimates* of the two population means. We are 95 percent confident only that that they fall between the respective intervals indicated in Figure 10.1. It follows that the difference between the sample means is also only an estimate of the difference between the two population means, and that we need a method of attaching confidence intervals to this estimate. This can be achieved by calculating the **standard error of the difference** from this formula (in which the subscripts 1 and 2 refer to samples 1 and 2, respectively):

$$\text{S.E.}_{\text{diff}} = \sqrt{\text{S.E.}_{\cdot 1}^2 + \text{S.E.}_{\cdot 2}^2}$$

Thus, rearranging:

$$\text{S.E.}_{\text{diff}} = \sqrt{\frac{s_1^2}{n_1} + \frac{s_2^2}{n_2}}$$

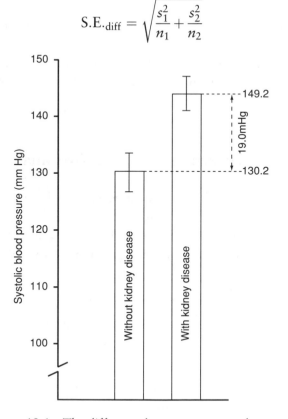

Figure 10.1   The difference between two sample means

For the blood pressure data (where $n = 30$ in each case):

$$\text{S.E.}_{\text{diff}} = \sqrt{\frac{19.4^2}{30} + \frac{21.2^2}{30}} = 5.25$$

Because the samples are regarded as large, we multiply the S.E.$_{\text{diff}}$ by the z-score of 1.96 to convert to an approximate 95 percent confidence interval. We are therefore 95 percent confident that the population mean difference $(\mu_1 - \mu_2)$ lies within $(\bar{x}_1 - \bar{x}_2) \pm 1.96\text{S.E.}_{\text{diff}}$:

$$
\begin{aligned}
\text{Estimated population mean difference} \quad &= 19.0 \pm (1.96 \times 5.25) \\
&= 19.0 \pm 10.29 \, \text{mmHg} \\
&= 29.29 \, \text{mmHgg (upper limit)} \\
&\text{and} \quad 8.71 \, \text{mmHg} \quad \text{(lower limit)}.
\end{aligned}
$$

This result is clearly more informative than the unprocessed sample mean difference of 19.0 mmHg.

Note that if the larger sample mean is nominated as $\bar{x}_1$ and the smaller as $\bar{x}_2$, then the sample mean difference is conveniently positive.

10.6   The difference between the means of two small samples

The rationale for establishing a confidence interval about the difference between means of two small samples is the same for that of large samples. The estimation of the standard error is a little more difficult, however. The expression for the standard error of the difference is:

$$\text{S.E.}_{\text{diff}} = \sqrt{\left[\frac{(n_1 - 1)s_1^2 + (n_2 - 1)s_2^2}{(n_1 + n_2 - 2)}\right]\left[\frac{(n_1 + n_2)}{(n_1 n_2)}\right]}$$

Although this is a rather cumbersome equation, the terms within it are familiar: they are simply the sample sizes and standard deviations of the samples being considered.

Example 10.5

Estimate the difference between the means of the two samples of systolic blood pressure measurements for which the sample statistics are given in Example 10.4 if the sample sizes are now 12 and 10 for patients without and with kidney disease, respectively. The difference between the sample means is 19.0 mmHg as before. Using the equation above (where n_1, s_1 and n_2, s_2 refer to *without* and *with* kidney disease, respectively):

$$\text{S.E.}_{\text{diff}} = \sqrt{\left[\frac{(11 \times 19.4^2) + (9 \times 21.2^2)}{(20)}\right]\left[\frac{(12 + 10)}{(120)}\right]}$$

$$= \sqrt{\frac{4139.96 + 4044.96}{20} \times 0.1833}$$

$$= \sqrt{75.01} = 8.66$$

Because the samples are small, the use of the z-score of 1.96 is not appropriate; t at $(n_1 + n_2 - 2)$ degrees of freedom is used in its place. For $P = 0.05$, t_{20} is 2.086. The confidence interval is therefore:

C.I. $= 19.0 \pm (2.086 \times 8.66)$
$= 19.0 \pm 18.06\,\text{mmHg}$
$= 37.06\,\text{mmHg (upper limit) and } 0.94\,\text{mmHg (lower limit).}$

This is, of course, a considerable extension of the confidence interval estimated for the larger samples of Example 10.4.

0.7 Estimating a proportion

Nurses and health care workers often express the frequency of an item in a sample as a proportion of the total (and then as a percentage – see Section 3.7). For example, 25 individuals from a sample of 200 in a community tested positive for sickle cell trait. The proportion is 25/200 $= 0.125$ (12.5%). This proportion is an estimate of the proportion in the population. Subsequent samples are, likewise, independent estimates of the same population proportion but, due to sampling error, are likely to be different from each other. In the same way that a set of sample means is distributed around a population mean, so too is a set of sample proportions distributed around a population proportion. The standard deviation of the distribution is similarly called the **standard error**. For practical purposes, a satisfactory estimate of the standard error from a sample is given by:

$$ \text{S.E.} = \sqrt{\frac{p(1-p)}{(n-1)}} $$

where p is the proportion of the nominated item, and n is the number of all items in the sample.

Example 10.6

Haemoglobin electrophoresis tests on a sample of 200 patients from a black Caribbean community revealed that 25 tested positive for sickle cell trait (i.e. are carriers of the gene for sickle cell disorder). Estimate the standard error and 95 percent confidence interval of the proportion of patients with the condition.

The proportion in the sample is $25/200 = 0.125$. Substituting in the formula:

$$ \text{S.E.} = \sqrt{\frac{0.125(1-0.125)}{(200-1)}} = 0.0234 $$

The limits of the confidence interval are obtained by multiplying the standard error by 1.96. The interval is:

$$ 0.125 \pm (1.96 \times 0.0234) = 0.125 \pm 0.0459 $$

That is, 0.171 (17.1%) upper limit and 0.079 (7.9%) lower limit. We are therefore 95 percent confident that in the population as a whole, the proportion of people with sickle cell trait is between 0.079 and 0.171 (7.9% and 17.1%).

Two points should be noted here. First, dramatic increases in sample size are required to reduce the confidence interval by an appreciable extent. Thus, increasing the sample size from 200 to 300 in Example

10.6 reduces the standard error from 0.0234 to 0.0191, a reduction of only 0.0043. Secondly, the formula for calculating the standard error of a proportion becomes unreliable when p (the proportion of the nominated item) is less than 0.1 or greater than 0.9.

0.8 The finite population correction

In Example 10.6 we considered a sample of 200 patients from a population of unknown, but certainly very large, size. If our population size is finite and known, and we have sampled more than 5 percent of it, we are able to apply a correction factor that improves the estimate of the proportion in the population. Such would be the case in our illustration of *simple random sampling* in Section 2.7, in which the entire population comprised 800 asthmatic patients (N) registered with a GP practice. A random sample of, say, 80 patients (10 percent) from this population would call for the application of the correction factor.

The formula for the Standard Error of a proportion applying the correction is similar to the one described in Section 10.6, but with an additional component in the square root term:

$$S.E. = \sqrt{\left(\frac{p(1-p)}{n}\right)\left(\frac{N-n}{N}\right)}$$

Example 10.7

In a sample of 100 from a population of 800 patients registered with a GP surgery, 74 patients have a family history of asthma. Estimate the proportion of patients in the study population of 800 who have a family history of asthma.

Proportion in the sample $= 74 \div 100 = 0.74$

Since we know that $N = 800$ and $n = 100$, we can substitute in the formula above:

$$S.E. = \sqrt{\left(\frac{0.74(1-0.74)}{100}\right)\left(\frac{800-100}{800}\right)} = 0.041$$

The 95 percent confidence interval is $\pm 1.96 \times$ S.E. $= 1.96 \times 0.041$ $= \pm 0.080$.

We are therefore 95 percent confident that the proportion of patients in the population who have a family history of asthma lies between 0.74 $\pm$ 0.080, that is, 0.82 (upper limit) and 0.66 (lower limit). Without the correction, the limits are 0.826 and 0.654, an extension of the confidence interval by a considerable 7.5 percent.

11 THE BASIS OF STATISTICAL TESTING

1.1 Introduction

In Section 9.3 we said that statisticians set arbitrary critical thresholds of probability. When an event occurs whose probability is estimated to be below a critical threshold, the outcome is said to be *statistically significant*. The critical values are: $P < 0.05$ (*significant*); $P < 0.01$ (*highly significant*); and $P =< 0.001$ (*very highly significant*).

Example 9.5 shows that we can use the properties of the normal curve to estimate probabilities. Thus, an observation obtained randomly from a normally distributed population and having a value deviating from the population mean (μ) plus or minus 1.96 standard deviations occurs in less than one trial in 20. That is to say, the probability of such an outcome is $P < 0.05$. If a single random observation *should* exceed this critical value, it is regarded as being statistically significant. The procedure for deciding if the outcome of a particular event is significant is called a **statistical test**. We now need to explain in more detail the basis of statistical testing.

1.2 The experimental hypothesis

The formulation and testing of hypotheses is the basis of experimental science, including medical science. A hypothesis is a proposed explanation for a state of affairs. A hypothesis is tested by experimentation, the outcome of which may provide evidence for the acceptance or rejection of the hypothesis. As an example of an experimental hypothesis we refer to the blood pressure measurements of patients with and without kidney disease described in Section 10.5:

The state of affairs:	Patients with kidney disease have, on average, higher blood pressure than those without.
Experimental hypothesis:	Kidney disease damages the blood capillaries in kidney tissue making it harder for the blood to circulate.
Possible experiment:	Examine microscopically histological sections of diseased and undiseased kidney tissue.

If it is found that diseased kidney tissue *does* damage the capillaries, the hypothesis is accepted for the time being. The outcome of the experiment does not *prove* that damaged capillaries are responsible for the elevated blood pressure (there might be other factors); it simply fails to provide evidence for doubting it. On the other hand, if microscopic examination of tissue sections failed to reveal capillary damage, the hypothesis would be rejected and an alternative proposed.

11.3   The statistical hypothesis

In our blood pressure example above, the description of the *state of affairs* is based on the observation that the mean blood pressure of a sample of patients with kidney disease is higher than that of a sample of patients without disease. Could there be any reason to doubt the validity of this difference? To examine this suggestion we nominate the mean blood pressure of the population of patients with kidney disease as μ_1 and that without disease as μ_2. Let us assume for the moment that there is *no* difference between the means of the two populations (with and without disease), and that $\mu_1 = \mu_2$. (This is actually called a null hypothesis, explained below.)

   If the population means are identical, we can superimpose the size frequency distribution of one population over the other, as we have done in Figure 11.1 (we have scaled the figure so that the two distributions have approximately the same height). Now, imagine that a small random sample is drawn from each population. Although each unit (a patient) is drawn randomly, it is possible, by chance, that one sample has more than a 'fair share' of larger blood pressure values. This is due to *sampling error*, described in Section 10.1. If, also by chance, the other

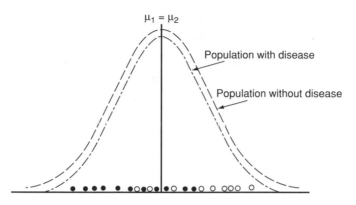

Figure 11.1 Size frequency distributions of systolic blood pressure measurements from populations of patients with and without kidney disease. •, random observations of diseased patients; o, random observations from patients without disease

sample has more than a 'fair share' of smaller values, then there could be a substantial difference between the means of the two samples, *even though they are drawn from populations with equal means.*

The likelihood of such a spurious event arising by chance diminishes rapidly as the number of observations in each sample increases. A statistical test is required to determine the probability that two samples of given size and with a particular mean difference could have been drawn from two populations with identical means. If the calculated probability is below the critical threshold, we accept the evidence and conclude that the means of the two samples are significantly different, and have been drawn from populations with different means. The basis of a statistical test that assumes there is *no* difference. This is called a **null hypothesis,** and is symbolized by H_0. An alternative to a null hypothesis is symbolized by H_1. Usually, an alternative hypothesis is simply that there *is* a difference, for example, $\mu_1 \neq \mu_2$.

By analogy with the experimental hypothesis of Section 11.2, we write:

State of affairs: The mean blood pressure of a sample of patients with kidney disease is higher than a sample without the disease.

Null hypothesis H_0:   Although the sample means are different, we assume for the moment that the two samples have been drawn from populations with equal means, i.e. $\mu_1 = \mu_2$. If H_0 is false, then $\mu_1 \neq \mu_2$.

Experiment (Statistical test):   A statistical test that computes the probability that two samples of their particular size and mean difference could have been drawn from two populations with identical means.

If the outcome of the test suggests that the probability of the two samples being drawn from two populations with identical means is too low to be acceptable, we reject H_0 and accept our alternative hypothesis, H_1. If this is the case, we may confidently proceed with our research to establish the cause of the difference, assured there is little risk that our evidence is based upon spurious sampling error.

11.4   Test statistics

The objective of each of the statistical tests that we shall describe is to produce a single number called a **test statistic**. The important feature about a test statistic is that its probability distribution is known, having been worked out by statisticians. z, whose probability distribution we outline in Section 9.7, is such a statistic. For example, we note in Section 9.7 that the probability (P) of an observation falling outside a z-value of ± 1.96 is less than 0.05; such an observation is considered *significant*. In small samples, the same consideration applies to the appropriate value of t.

In each case, a statistical test produces a value for its test statistic, and the task is to determine whether the value exceeds some probability threshold that suggests rejection of a Null Hypothesis. Fortunately, published tables of the threshold values of each test statistic are readily available and, as we shall show, are easy to use.

11.5   One-tailed and two-tailed tests

In Section 9.8 we distinguish between one-tailed tests and two-tailed tests. Now we may relate these to the idea of hypothesis-testing.

In our example in Section 11.3 we establish a null hypothesis that two samples are drawn from two populations with identical means. We make no prediction, if H_0 is false, as to which mean is the larger than the other – only that they are different:

$$H_0 : \mu_1 = \mu_2$$
$$H_1 : \mu_1 \neq \mu_2$$

If the outcome of the test is such that we are obliged to reject H_0 and accept H_1, we are then entitled to assume that the sample with the larger mean has been drawn from a population with a larger mean. This conclusion is only reached *after* the test (that is to say, *a posteriori*). A test like this which makes no prediction as to which mean is the larger of the two, should they prove to be different, is called a **two-tailed** test.

Sometimes it is possible to predict in advance that if two samples are not drawn from populations with equal means, then a *nominated* sample has been drawn from a population with a larger mean or, alternatively, has been drawn from a population with a smaller mean. The null hypothesis is the same, but the alternative is different. If the population predicted to have the larger mean is nominated μ_1 then:

$$H_0 : \mu_1 = \mu_2$$
$$H_1 : \mu_1 > \mu_2$$

Because a one-tailed test is less stringent than a two-tailed test, considerable caution should be taken before using it. Moreover 'significant' results can be obtained with smaller sample sizes. Observers who persuade themselves that a test is one-tailed in order to obtain a 'result' with a smaller sample are *definitely cheating*!

In summary, a one-tailed test should only be used when there is an *a priori* reason to predict a directional influence in the data; moreover, the decision to use a one-tailed test *must* be made before the data are analysed. If there is doubt – use a two-tailed test; an outcome which is statistically significant in a two-tailed test is also significant in a one-tailed test.

There is no difference between the execution of a one- tailed test and a two-tailed test; there is simply a lower threshold of significance in a one-tailed test.

1.6   Hypothesis testing and the normal curve

We may now relate the principles outlined in this chapter to the statistical tests undertaken in Chapter 9. Turn again to Example 9.5. There we ask if it is likely that a single observation of 2.0 kg is drawn randomly from a population for which $\mu = 3.8$ and $\sigma = 0.5$. Our null hypothesis H_0 is that the observation *is* drawn from such a population (H_1 is that it is not). The test involves the computation of a test statistic z. The computed value of z (3.6) exceeds 1.96 which is the value corresponding to the critical probability threshold of $P = 0.05$. We therefore reject H_0 and accept H_1, and conclude that the observation is probably not drawn from that population. In a one-tailed test the critical value of z is 1.65. Figure 11.2 illustrates how the normal curve is used in hypothesis testing.

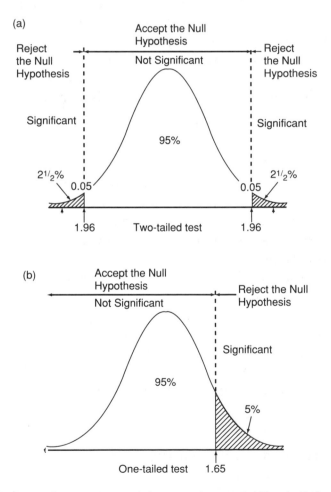

Figure 11.2　Hypothesis testing and the normal curve. (a) Two-tailed test; (b) one-tailed test

11.7   Type 1 and type 2 errors

When the threshold for rejection of H_0 is set at $P = 0.05$, an investigator is said to *accept the 0.05 (or 5 per cent) level of significance*. This means that in tests where the computed value of the test statistic is equal to, or barely exceeds, the critical value, the decision to reject H_0 is probably correct 19 times out of 20, or 95 times out of 100. It follows that 5 times out of 100 there is a risk of rejecting H_0 when it is *true*. When H_0 is rejected and it is actually true, we refer to a **type 1 error** having been committed. How can the risk of a type 1 error be reduced? Simply by setting the level of acceptance at a more rigorous standard, for example at the 1 in 100 times level of significance ($P = 0.01$).

It will be appreciated, however, that the analyst faces a 'swings and roundabouts' situation. The opposite of the case just outlined is referred to as a **type 2 error**, that is *not* rejecting H_0, when in fact it should be rejected.

Thus, as the probability of making a type 1 error is reduced, the probability of making a type 2 error is increased.

In most statistical analyses the aim is usually to limit the probability of committing type 1 errors, thus erring on the side of caution. In practice, the calculated value of a test statistic often exceeds the tabulated critical value at $P = 0.05$, in which case we reject H_0 at $P < 0.05$ and the risk of error is accordingly reduced.

11.8   Parametric and non-parametric statistics: some further observations

In Section 2.15 brief reference is made to the distinctions between parametric and non-parametric statistics. It is noted that the conditions relating to the use of parametric tests are more rigorous than those that apply to non-parametric tests. These distinctions are now elaborated upon in Table 11.1.

We may wish to test nominal and ordinal level data, in which case non-parametric tests are often very suitable. Non-parametric tests usually employ simple formulae, and avoid the rather tedious computations of variances, sums of squares, etc. that are often required in parametric tests (though, of course, computers will do much of the hard work for us!). Furthermore, they require no data transformation. On the other hand, because non-parametric tests may often be used with

Table 11.1 Distinctions between non-parametric and parametric tests

Non-parametric tests	Parametric tests
May be used with actual observations, or with observations converted to ranks	Are used only with actual observations or their transforms
May be used with observations on nominal, ordinal and interval scales	Generally restricted to observations on interval scales
Compare medians	Compare means and variances
Do not require data to be normally distributed or to have homogeneous variance; i.e. they are 'distribution free'.	Require data to be normally distributed and to have homogeneous variances
Are suitable for data which are 'counts'	Counts must usually be transformed
Are suitable for derived variables, e.g. rates and proportions	Derived data may first have to be transformed

smaller samples, do not use *all* the data (that is **ranks** rather than the actual observations are used) and are more 'permissive', they may be less powerful than a corresponding parametric test.

11.9   The power of a test

Statisticians sometime refer to the **power of a test.** It is a measure of the likelihood of a test reaching the correct conclusion, i.e. rejecting H_0 when it should be rejected. Non-parametric tests are generally regarded as being less powerful than parametric equivalents, because they are less rigorous in their conditions of use. It must be emphasized, however, that parametric tests are more powerful than non-parametric only when the assumptions governing their use hold true. If an assumption does not hold true (e.g. the data do not fit a normal distribution), it is not just a question of reduction of power; the whole validity of the test is destroyed, and there may be the risk of considerable error.

There is one safe rule: *If there is doubt as to whether a particular set of data satisfies the assumptions made in the use of a parametric test, then a non-parametric alternative should be used.*

12 ANALYSING FREQUENCIES

12.1 The chi-square test

Nurses and other health care workers often count and classify things on nominal scales such as sex, age-group, blood group type, ethnic origin, and so on. Statistical techniques, which analyse frequencies are, therefore, very useful. The classical method of analysing frequencies is the chi-square test (pronounced 'ky' as in 'sky').

Chi-square tests are variously referred to as tests for **homogeneity, randomness, association, independence** and **goodness of fit**. This array is not as alarming as it might seem at first sight. The precise applications will become clear as you study the examples. In each application the underlying principle is the same. The frequencies we *observe* are compared with those we *expect* on the basis of some null hypothesis. If the discrepancy between the observed and expected frequencies is great, then the value of the calculated test statistic will exceed the critical value at the appropriate number of degrees of freedom. We are then obliged to reject the null hypothesis (that is, that the 'observed' and 'expected' are in agreement) in favour of some alternative.

The test involves computing a test statistic that is compared with a chi-square (χ^2) distribution that has been worked out and tabulated by statisticians (see Appendix 3). In Appendix 3 the distribution is restricted to the critical values at the significance levels we are interested in, namely $P = 0.05$ and $P = 0.01$.

The mastery of the method lies not in so much in the computation of the test statistic itself, but in the calculation of the expected frequencies.

12.2 Calculating the test statistic

The simplest arithmetical comparison that can be made between an observed an expected frequency is the difference between them. In the test, the difference is squared and divided by the expected frequency. Thus:

$$\chi^2 = \frac{(O - E)^2}{E}$$

where O is an observed frequency and E is an expected frequency.

In practice, a series of observed frequencies is compared with corresponding expected frequencies resulting in several components of χ^2 all of which have to be summed. The general formula for χ^2 is therefore:

$$\chi^2 = \Sigma \frac{(O - E)^2}{E}$$

To illustrate the calculation of the χ^2 statistic we choose a 'desk top' example.

Example 12.1

On your calculator you should find a key that generates a random number (usually marked 'Ran#'). After a few trials, a student harbours a suspicion that the integers generated are not truly random. To test this, 100 numbers are generated and displayed as a frequency distribution (the first digit only of any number is selected):

Random number, x	0	1	2	3	4	5	6	7	8	9	
Frequency of occurrence, f	10	7	10	6	14	8	11	11	12	11	($n = 100$)

The problem now is to work out how many times we should *expect* each integer to occur if they are indeed being drawn randomly. This example permits two possible approaches to the problem, namely the *theoretical* and the *empirical*. First, independent integers generated randomly have, by definition, an equal probability of being drawn. Since one-digit numbers are selected (0–9, i.e. ten possibilities) the probability of any one is 0.1. Since we know (Section 9.4) that:

Expected frequency = probability of occurrence × sample size

we can compute the expected frequency of each as:

$$= 0.1 \times 100$$
$$= 10$$

The expected frequency of each integer is therefore 10. Alternatively, if our sample size, n, is 100, the null hypothesis predicts there will be a *homogeneous* (that is, evenly spread) and *independent* distribution of numbers. Since there are 10 possible integers, a homogeneous distribution is $100 \div 10 = 10$ of each. Thus, the expected frequency of each is again 10. The test can be either a test for *randomness* or a test for *homogeneity*, or *independence*, according to your point of view.

We can now write:

Random number, x:	0	1	2	3	4	5	6	7	8	9
Observed frequency, O:	10	7	10	6	14	8	11	11	12	11
Expected frequency, E:	10	10	10	10	10	10	10	10	10	10

Having now worked out the expected frequencies, it is a simple, matter to calculate the test statistic:

$$\chi^2 = \frac{(10-10)^2}{10} + \frac{(7-10)^2}{10} + \frac{(10-10)^2}{10} + \frac{(6-10)^2}{10} + \frac{(14-10)^2}{10}$$
$$+ \frac{(8-10)^2}{10} + \frac{(11-10)^2}{10} + \frac{(11-10)^2}{10} + \frac{(12-10)^2}{10} + \frac{(11-10)^2}{10}$$

$$\chi^2 = 0 + 0.9 + 0 + 1.6 + 1.6 + 0.4 + 0.1 + 0.1 + 0.4 + 0.1$$
$$\chi^2 = 5.2$$

Having calculated the test statistic, the next step is to evaluate the number of degrees of freedom (df). The rule is simple. It is the number of categories (a) less one. In this example, there are therefore 9 df. We now have to determine whether the calculated value of 5.2 at 9 df exceeds the critical value at a chosen level of significance. This is decided by reference to Appendix 3. Turn to Appendix 3 and run down the left column until the number of degrees of freedom (9 in this case) is reached. Looking across the table at that point we find two values: 16.92 under the 0.05 level of significance and 21.67 under the 0.01 level. Our value of 5.2 is smaller than either of these. We accept the null hypothesis that the observed frequencies agree with the expected frequencies. We are justified in concluding that the numbers *are* homogeneous and probably *were* generated independently and randomly. We can express the result in the form: 'A statistical test shows that the frequency of each integer is consistent with having been generated randomly, $\chi_9^2 = 5.2$ (NS)'. The subscript 9 below the χ^2 is a shorthand way of indicating the 9 df, and NS stands for *not significant*, that is, no significant difference between observed and expected frequencies.

12.3   A practical example of a test for homogeneous frequencies

Example 12.2

A sample of 195 white male blood donors are classified into blood group types. Each type is a truly nominal category, and the sample of donors is assumed to be random. The results are:

	Blood group type			
A	B	AB	O	*Total*
58	34	9	94	195

Whilst at first sight it certainly does not *appear* that the blood group types are spread evenly between the categories, the question to be asked is: 'Is the distribution between the categories in fact homogeneous? Alternatively, can the observed variation be accounted for by chance scatter (sampling error)?' In either case the statistical hypotheses are:

H_0: the observed frequencies are homogeneous and the apparent departure from homogeneity is merely due to sampling error or scatter.

H_1: the observed frequencies depart from those expected of a homogeneous distribution by an amount that cannot be explained by sampling error.

If H_0 is true, we expect that the 195 observations should be distributed evenly between the four categories, that is $195 \div 4 = 48.75$ (note that the 'expected frequencies' do not have to be a whole number).
    The observed and expected frequencies are:

A	B	AB	O
58	34	9	94
48.75	48.75	48.75	48.75

Compute the test statistic exactly as in Example 12.1:

$$\chi^2 = \Sigma \frac{(O - E)^2}{E} = 1.75 + 4.46 + 32.4 + 42.0 = 80.61 \text{ with } (4 - 1) = 3 \text{ df.}$$

From Appendix 3 we see that our calculated test statistic exceeds the critical value at both $P = 0.05$ and $P = 0.01$. We conclude that if the frequencies in the four categories are, in fact, distributed homogeneously in the population, then a discrepancy of this magnitude would happen by chance in a sample on fewer than 1 occasion in 100. We therefore reject H_0 in favour of H_1, and record the outcome: 'There is a statistically highly significant departure from homogeneity between the four categories, $\chi^2_3 = 80.61$, $P < 0.01$'. We may infer that there are significantly more individuals in the population exhibiting blood group type O and fewer with type AB. Indeed, as we shall see in a later example (Example 12.4), there is good reason for supposing that there is an underlying reason for this state of affairs.

12.4 One degree of freedom – Yates' correction

Where there are only two categories in a distribution there is only one degree of freedom. In this case, the calculated value of the test statistic is too high unless we make a correction called **Yates' Correction for Continuity**. This involves subtracting 0.5 from the numerator (i.e. the top portion of the formula) of each component of the chi-square formula before squaring. The subtraction is made to the *absolute value* of the difference $(O - E)$, that is, any minus sign is ignored. This is written as $(|O - E|)^2$, where the vertical bars on each side of $(O - E)$ mean *absolute value*.

Example 12.3

On the first day of the millennium a maternity ward announced the birth of 16 babies, of which 12 were girls and 4 were boys. The local press commented on the seemingly large discrepancy between the gender balance. Does this, however, constitute a statistically significant departure from 1:1 (50:50), or can the discrepancy be reasonably accounted for by chance?

The null hypothesis H_0 is that in the population the gender ratio *is* unity, that is, the ratio of females to males is 1:1. If this is the case we would expect eight males and eight females in the sample of 16. Gender is a truly nominal category. The χ^2 test for continuity is therefore appropriate. Because there is 1 df we apply Yates' correction:

- *Without* the correction, the value of the test statistic is $[(12 - 8)^2/8]$ $+[(4 - 8)^2/8] = 4.0$. The value exceeds the tabulated critical value of 3.84 at 1 df and is significant at $P = 0.05$.

- *With* the correction it is $[(|12 - 8| - 0.5)^2/\ 8] + [(|4 - 8| - 0.5)^2/8]$ $= (1.531 + 1.531) = 3.062$. This is *less* than the tabulated critical value of 3.84 at $P = 0.05$.

We conclude that there is *no* significant departure of the sex ratio from 1:1. Notice how in this particular example the application of the correction alters the outcome of the test.

12.5   Goodness of fit tests

The procedure for executing a **goodness of fit** test is exactly the same as for a homogeneity test, except that the expected frequencies are generated according to some established or predicted rule or theory. For example, considering our previous blood group type example (Example 12.2), it is well-known that blood group types are not found evenly in a population but are distributed according to underlying laws of genetics, e.g. the Hardy Weinburg Law.

Example 12.4

Geneticists predict that in a particular community the ratio of blood group types A, B, AB and O should exist in the ratio 10 : 4 : 1: 15 (for illustrative purposes, we have simplified this). Do the frequencies of the 195 donors in Example 12.2 conform to this ratio?
We recall that the observations were:

Blood group type				
A	B	AB	O	*Total*
58	34	9	94	195

We now have to work out how many of each blood group type we should *expect* if the ratio is true. The 'sum of the ratios' is $10 + 4 + 1 + 15 = 30$. So the expected frequencies are found by dividing the total (195) by 30 and multiplying by the expected ratio factor. Thus, for Type A we expect $195 \div 30 \times 10 = 65$. The results for all types are summarized below:

Blood group type	A	B	AB	O
Expected ratio	10	4	1	15
Observed frequency	58	34	9	94
Expected frequency	65	26	6.5	97.5

We can already see that there is closer agreement between the 'observed' and 'expected' frequencies. To decide if the 'fit' is close enough to accept the prediction, we calculate the χ^2 statistic in the usual way:

$$\chi^2 = 0.754 + 2.46 + 0.96 + 0.126 = 4.3$$

Consulting Appendix 3 at 3 df, our calculated value of 4.3 is *less* than the tabulated value of 7.81. We therefore *accept* the null hypothesis, and conclude that the observed and expected frequencies are in agreement (i.e. there is a 'good fit'), and there is no evidence to contradict the predictions of the geneticists.

12.6   The contingency table – tests for association

In all of the previous examples of chi-square analysis, observed frequencies are distributed between categories in *one row*. When this is the case we refer to a **one-way classification**. Sometimes, however, two nominal level observations are obtained from a sampling unit. Thus, we may record the hair colour of an individual and sex; the ethnic background of an individual and location; a disease category and age, and so on. In such cases, frequencies may be arranged in *two or more rows*, and we refer to a **two-way classification**. Tables of these data are called **contingency tables**. They allow the investigation of *association* between variables. The simplest type of contingency table is one that has only two nominal categories of each variable. It is called a **2 × 2 table**. An example follows.

Example 12.5

A survey is conducted to investigate associations between decedent characteristic (mode of death) and location. The numbers of cases of patients dying of heart disease and cancer at home or in hospital are shown in Table 12.1. Cells in the table are conventionally labelled *a, b, c, d* and the row totals, column totals and grand totals are included.

   At first sight it appears that there is a higher proportion of patients with cancer who die at home, while a higher proportion with heart disease die in hospital. If we can demonstrate that the proportional differences cannot be accounted for by sampling error (scatter, random chance), then we could assert that there is an association between the variables. Chi-square contingency analysis is appropriate to the problem.

   To apply the χ^2 test we need to calculate the expected frequency in each cell. This task is complicated by the fact that we are in fact dealing with two null hypotheses:

Table 12.1   Frequency of patient deaths

	Cancer	Heart disease	Totals
Home	a 84	b 27	a + b 111
Hospital	c 16	d 50	c + d 66
Totals	a + c 100	b + d 77	a + b + c + d 177

1. The ratios of the frequencies in both vertical columns do not depart from the overall vertical ratio (i.e. 84:16 and 27:50 do not depart from 111:66).
2. The ratios of the frequencies in both horizontal rows do not depart from the overall horizontal ratio (i.e. 84:27 and 16:50 do not depart from 100·77).

To calculate the expected frequency for any single cell, we multiply the total for its column by the total for its row and divide by the grand total. Thus, the expected frequencies for cell a: 'Cancer at home' is $[(a + c) \times (a + b)] \div (a + b + c + d) = (100 \times 111 \div 177) = 62.7$.

The table may now be written out (in abbreviated form) showing the observed and expected frequencies in each cell.

Cell a	*Cell b*
$O = 84$	$O = 27$
$E = 100 \times 111 \div 177 = 62.7$	$E = 77 \times 111 \div 177 = 48.3$
Cell c	*Cell d*
$O = 16$	$O = 50$
$E = 100 \times 66 \div 177 = 37.3$	$E = 77 \times 66 \div 177 = 28.7$

(Check that the sum of the expected frequencies equals the sum of the observed frequencies, $a + b + c + d = 62.7 + 48.3 + 37.3 + 28.7 = 177$.)

With the four observed frequencies and their respective expected frequencies, the individual components of the test statistic can now be calculated. Before we do that, it is advisable to determine the number of degrees of freedom. The rule for determining the degrees of freedom in a contingency table is:

Degrees of freedom = number of columns (c) minus 1, multiplied by the number of rows (r) minus 1
That is: $(c - 1)(r - 1)$. In our example this is $(2 - 1)(2 - 1) = 1$.

A 2×2 contingency table, therefore, has only one degree of freedom. With the condition described in Section 12.4 in mind, we must apply Yates' correction to each cell in the table. Readers may note that a 2×2 table is the *only* one in which the correction needs to be applied.

Taking the four cells, respectively, the test statistic is the sum of:

Cell a	*Cell b*
$(\lvert 84 - 62.7 \rvert - 0.5)^2/62.7 = 6.90$	$(\lvert 27 - 48.3 \rvert - 0.5)^2/48.3 = 8.96$

Cell c	*Cell d*
$(\lvert 16 - 37.3 \rvert - 0.5)^2/37.3 = 11.60$	$(\lvert 50 - 28.7 \rvert - 0.5)^2/28.7 = 15.07$

$$\chi^2 = 6.90 + 8.96 + 11.60 + 15.07$$
$$= 42.53$$

Consulting Appendix 3, we find that the calculated value of 42.53 at 1 df greatly exceeds the tabulated value at both 5 percent and 1 percent (0.05 and 0.01) levels of significance. There is, therefore, a statistically highly significant association between the variables. In particular, cancer patients tend to die at home; heart disease patients tend to die in hospital.

There is wide application for 2×2 contingency tables in statistical analysis. Note that nominal categories may be represented by 'yes' and 'no', or 'with' and 'without'.

Example 12.6

An accident and emergency department conducted a survey of patients who presented with head injuries following cycling accidents. In particular, the analyst was interested in discovering if there was an association between injuries to the head and the use of a helmet. The data are presented in Table 12.2.

Notice that for each variable, the nominal categories are 'yes' and 'no'. We suggest that you execute the contingency analysis exactly as shown in Example 12.5 (remember to apply Yates' correction!), and confirm that $\chi_1^2 = 46.3$, showing a highly significant association between the occurrence of head injuries and wearing a helmet. In this

Table 12.2 Head injuries and helmets

Head injury	Wearing helmet		Total
	Yes	No	
Yes	15	155	170
No	125	202	327
Total	140	357	497

case, the association is *negative*, there are proportionately *fewer* head injuries to those who wear a helmet.

Example 12.7

We say in Section 6.4 that the significance of the outcome of case control studies in epidemiology can be tested by chi-square tests. We repeat here the table of data from Example 6.4, in which the relationship between smoking and cancer of the mouth and pharynx is examined.

	Smokers		
	Yes	No	Totals
Cases	457	26	483
Controls	362	85	447
Totals	819	111	930

We suggest that you execute the contingency analysis exactly as shown in Example 12.5 (remember to apply Yates' correction!), and confirm that $\chi_1^2 = 39.8$. Clearly, there is an epidemiological association between smoking and the cancer.

12.7   The 'rows by columns' ($r \times c$) contingency table

Where there are more than two nominal categories in a two-way classification, a contingency table may have several rows and columns in it. If there are r rows and c columns, there are $r \times c$ cells in the table. The procedure for working out the expected frequencies and calculating the test statistic is similar to that for a 2×2 table, except that Yates' correction is not applied. To illustrate this method, we choose a 2×3 table.

Example 12.8

In a study of the implications of smoking habits for heart disease, a survey of 232 women of three ethnic origins revealed the following results (Table 12.3).

Table 12.3   Numbers of women smokers and non-smokers

	Whites	Blacks	Asians	Totals
Non-smokers	52	55	66	173
Smokers	32	21	6	59
Totals	84	76	72	232

It appears that white women have the highest proportion (32/84) of smokers. Is this observation 'significant', or could it be attributed to sampling error? Chi-square contingency analysis is appropriate to the problem. The steps in the analysis are as follows:

1. Calculate the six individual frequencies that would be expected if H_0 is true, that is, there is no association between any of the categories. By way of example, the expected frequency of 'white non-smoker' is $(84 \times 173) \div 232 = 62.64$ (column total $\times$ row total $\div$ grand total).

2. Calculate all the individual components of χ^2 from $(O - E)^2/E$. The results of the procedure thus far are shown in Table 12.4.

3. Sum all the individual values of $(O - E)^2/E$. The result is 18.33. This is the test statistic.

4. Determine the degrees of freedom from $(c - 1)(r - 1)$. This is $2 \times 1 = 2$.

5. Consult Appendix 3 at this number of degrees of freedom, and decide if the test statistic calculated in step (3) exceeds the critical value at $P = 0.05$ or $P = 0.01$. The critical value at $P = 0.01$ at 2 df is 9.21. The calculated value exceeds this.

6. Express the result in the form: 'There is a statistically highly significant association between smoking behaviour and ethnic background in females $\chi^2 = 18.33$, $P < 0.01$'.

To decide where the particular associations lay, that is, which ethnic groups are associated with which smoking habit, we examine the individual components of χ_2^2 in Table 12.4. Inspecting the table, we find the largest individual values are: 8.27 (Asian, smoking) and 5.24 (White, smoking). Looking at those cells more closely, we see that there are fewer Asian smokers than expected, and more White. We conclude that

Table 12.4 Calculation of expected frequencies

	Variable A: Ethnic background		
	Whites	Blacks	Asians
Non-smokers			
O:	52	55	66
E:	62.6	56.7	53.7
$(O-E)^2/E$	1.79	0.051	2.82
Smokers			
O:	32	21	6
E:	21.4	19.3	18.3
$(O-E)^2/E$	5.25	0.15	8.27

white females are proportionately heavier smokers (positively associated), and Asians are proportionately the least heavy smokers (negatively associated) in the population being studied.

12.8 Larger contingency tables

Contingency tables can, in principle, have any number of rows and columns. However, with very large contingency tables, interpreting the result in terms of pin-pointing the particular associations becomes more difficult. Nevertheless, tables with five or six columns or rows are not uncommon, and we illustrate these with two further examples.

Example 12.9

Table 12.5 is a 3×3 contingency table showing severity of depression and suicidal intent in a sample of 500 psychiatric patients.

We have calculated the 'expected frequencies' (in brackets) on the basis on a null hypothesis that assumes no association between the variables. We suggest you follow the procedure described in Example 12.8, and confirm that the expected frequencies are correct, and that the chi-square statistic = 71.45 with 4 df. This value exceeds that tabulated at $P = 0.001$, and therefore H_0 is rejected. There is a strong association between suicidal intent and severity of depression.

Table 12.5   Depression and suicidal intent (expected frequencies in brackets)

	Not depressed	Moderately depressed	Severely depressed	Total
Attempted suicide	26 (50.13)	39 (33.07)	39 (20.80)	104
Contemplated or threatened suicide	20 (35.67)	27 (23.53)	27 (14.80)	74
Neither	195 (155.20)	93 (102.40)	34 (64.40)	322
Total	241	159	100	500

Example 12.10

In Section 4.7 we describe the distribution of the location of accidents between males and females in the form of 'pie graphs'. We are now able to decide if the apparent differences revealed in the graphs are 'statistically significant', or merely the product of chance. The data are reproduced in Table 12.6.

We suggest you follow the procedure described in Example 12.8, and verify that our expected frequencies (in brackets) are correct, and the chi-square statistic $= 266.64$ at 5 df and greatly exceeds the tabulated value at $P = 0.001$. H_0 is consequently rejected, and we conclude that there is a strong association between gender and location of accident. In particular, it appears that accidents among males are especially associated with the workplace, and those among females are associated with the home and garden.

Table 12.6   Gender differences in locations of accidents (expected frequencies in brackets)

Gender	Location of accident						Total
	Home & garden	Road	Work place	School or college	Sports area	Other	
Males	531	322	512	152	190	208	1915
	(695)	(349)	(380)	(137)	(136)	(218)	
Females	653	272	136	82	41	163	1347
	(489)	(245)	(268)	(97)	(95)	(153)	
Total	1184	594	648	234	231	371	3262

12.9 Advice on analysing frequencies

1. All versions of the chi-square test compare the agreement between a set of observed frequencies and those expected if some null hypothesis is true. They are all, in a sense *goodness of fit* tests, although this title is usually restricted to those in which the expected frequencies are estimated according to some underlying theory.

2. As objects (items) are counted they should be assigned to nominal categories. Remember that 'yes' and 'no' constitute such categories. Unambiguous intervals on a continuous scale may be regarded as *nominal* for the application of the tests. For example, successive days, weeks or months might comprise such a scale, or contiguous bands on a blood pressure scale. The 'degree of depression' of Example 12.8 is a further example.

3. The sample size, that is, the grand total of all observed frequencies (n), should be such that all *expected* frequencies exceed 5. In marginal cases, this can sometimes be achieved by collapsing cells and aggregating the respective observed and expected frequencies. Some flexibility in interpreting this rule is allowed. Most statisticians would not object to some of the expected frequencies being below 5, provided that no more than one-fifth of the total number of expected frequencies is below 5, and none are below 1.

4. Apply Yates' correction in the chi-square test when there is only one degree of freedom, that is, when there are two categories in a 'one-way' test and in 2×2 contingency tables.

13 MEASURING CORRELATIONS

13.1 The meaning of correlation

Many variables in nursing and health care are related; examples include the weight of a baby and its age; the weight of a child and its height; the age of a person and blood pressure; the diameter on an ulcer and the time it takes to heal. Relationships or associations between variables such as these are referred to as **correlations**. Correlations are measured on ordinal or interval scales.

When an increase in one variable is associated by an increase in another, the correlation is said to be **positive** or **direct**. The height and weight of a child is usually positively correlated. When an increase in one variable is accompanied by a decrease in the other the correlation is said to be **negative** or **inverse**. We would anticipate that reaction rate is inversely correlated with the number of units of alcohol consumed.

The fact that variables are associated or correlated does not necessarily mean that one *causes* the other. Height and weight may be correlated in a population, but one variable cannot be said to *cause* the other: both are undoubtedly related to some underlying genetic factor. In common usage, the word 'correlation' describes any type of relationship between objects and events. In statistics, however, correlation has a precise meaning; it refers to a quantitative relationship between two variables measured on ordinal or interval scales.

13.2 Investigating correlation

Bivariate observations of variables measured on ordinal or interval scales can be displayed as a scattergram (Figures 4.7 and 13.1). Just as a simple dot plot gives both a useful indication of whether a sample of observations is roughly symmetrically distributed about a mean, and the extent

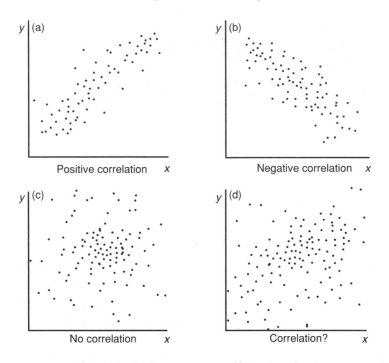

Figure 13.1   Scattergrams of bivariate data

of the variability, a scattergram gives an impression of correlation. Figure 13.1(a) shows a clear case of a positive correlation, whilst Figure 13.1(b) shows an equally clear case of a negative correlation. Figure 13.1(c) shows no correlation, but what about Figure 13.1(d)? It is not easy to be certain about this.

Subjective examination of a scattergram must be replaced by a statistical technique which is more precise and objective. The statistic that provides an index of the degree to which the two variables are related is called the **correlation coefficient**. The statistic is calculated from sample data, and is the estimator of the corresponding population parameter.

The numerical value of the correlation coefficient, r, falls between two extreme values: $+1$ (for perfect positive correlation) and -1 (for perfect negative correlation). A perfect correlation exists for interval data when all the points in a scattergram fall on a perfectly straight line. For ordinal (rank) data, a perfect monotonic relationship exists when all the values are successively ascending or descending, not necessarily on a perfect straight line (Figure 13.2).

Perfect or near perfect correlations (positive or negative) are virtually non-existent in the health-sciences: they are the privilege of the physicist!

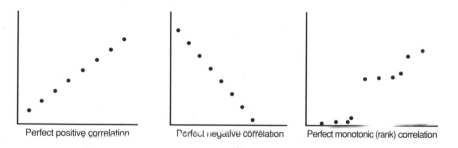

Figure 13.2   Perfect correlations

An example of a perfect correlation is described in Section 14.4. A correlation coefficient of 0, or near 0, indicates lack of correlation.

Correlation coefficients can be calculated by both parametric and non-parametric methods. A parametric coefficient is the **Product Moment Correlation Coefficient**. It is used only for interval scale observations (normally 'measurements'), and is subject to more stringent conditions than non-parametric alternatives. Of the various non- parametric coefficients, the **Spearman Rank Correlation Coefficient** is among the most widely used, and is the one we illustrate. It is appropriate for observations based on ordinal (rank) scales as well as interval scales.

3.3   The strength and significance of a correlation

In the previous section we said that all values of a correlation coefficient fall on a scale with limits −1 to +1. The closer the value of a coefficient is to −1 or +1, the greater is the strength of the correlation, whilst the closer it is to 0 the weaker it is. As a rough and ready guide to the meaning of the coefficient, Table 13.1 offers a descriptive interpretation.

Sometimes an apparently strong correlation may be regarded as *not statistically significant*, whilst a weak correlation may be *statistically*

Table 13.1   The strength of a correlation

Value of the coefficient r (positive or negative)	Meaning
0.00 to 0.19	A very weak correlation
0.20 to 0.39	A weak correlation
0.40 to 0.69	A modest correlation
0.70 to 0.89	A strong correlation
0.90 to 1.00	A very strong correlation

highly significant. We must resolve this apparent paradox. Look again at Figure 13.1(c), that we say shows no correlation. Imagine that the points in the scattergram represent a population of observations from which individual points can be nominated at random. The first two points are bound to be in perfect alignment, because a straight line can connect *any* two points. We should not be tempted, however, into thinking in terms of a correlation. It is not improbable that a third point, nominated randomly, could be in rough alignment with the other two, giving an impression of correlation. However, the more points that are nominated the less and less likely it becomes that a chance correlation can be maintained. When *all* the points in the population have been selected, we can measure the absolute correlation coefficient *parameter* ρ (rho). Any sample removed from the population estimates ρ. Large samples give reliable estimates and small samples give less reliable estimates. If the value of ρ is low, that is, a weak correlation in the population, large samples give good estimates and are *statistically significant*. On the other hand, even if ρ should be large, a small sample is nevertheless a poor estimate which may not be statistically significant. The point to bear in mind is that larger samples do not *strengthen* a weak correlation, but they do reduce the likelihood of a spurious correlation arising by chance or sampling error.

13.4   The product moment correlation coefficient

The Product Moment Correlation Coefficient is a parametric statistic that is appropriate when observations are measured on interval scales from randomly sampled sampling units. It is assumed that *both* variables are approximately normally distributed, that is to say, *bivariate normal*. This can be checked from a scattergram of the data. The broad outline of points in a scattergram of bivariate normal data is roughly circular or elliptical. The circle becomes drawn out into an ellipse as *r* increases in value.

The formula for the Product Moment Correlation Coefficient is given below. The formula may look rather fearsome, but all the terms within it are easily obtained from your scientific calculator:

$$r = \frac{n\Sigma xy - \Sigma x \Sigma y}{\sqrt{\left[n\Sigma x^2 - (\Sigma x)^2\right]\left[n\Sigma y^2 - (\Sigma y)^2\right]}}$$

Table 13.2 Chest circumference and birth weight of 10 babies. The bottom line shows the respective column totals

Chest circumference (cm) x	Weight (kg) y	x^2	y^2	xy
22.4	2.00	501.76	4.00	44.8
27.5	2.25	756.25	5.06	61.88
28.5	2.10	812.25	4.41	59.85
28.5	2.35	812.25	5.52	66.98
29.4	2.45	864.36	6.00	72.03
29.4	2.50	864.36	6.25	73.5
30.5	2.80	930.25	7.84	85.4
32.0	2.80	1024.0	7.84	89.6
31.4	2.55	985.96	6.50	80.07
32.5	3.00	1056.25	9.00	97.5
Σ 292.1	24.8	8607.69	62.42	731.61

Example 13.1

Babies that have low birth weight are especially vulnerable, and need to be identified quickly. In developing countries, logistic problems and availability of appropriate apparatus may prevent the weighing of every new-born child. However, if low birth weight could be predicted by means of a simpler method (e.g. using just a tape measure), then vulnerable neonates may be detected without directly weighing. S. K. Bhargava *et al.* (1985), working in India, measured a number of variables of new-born babies, including weight, head circumference and chest circumference. By way of example, we illustrate the usefulness of correlation by reference to their chest circumference data.

A sample of new-born babies was both weighed and measured for chest circumference. The results are shown in Table 13.2. Is there a strong correlation between birth weight and chest circumference?

Summarizing the data from Table 13.2:

$$\Sigma x = 292.1 \qquad \Sigma y = 24.8 \qquad n = 10$$
$$(\Sigma x)^2 = 292.1^2 \qquad (\Sigma y)^2 = 24.8^2 \qquad \Sigma xy = 731.6$$
$$= 85322.41 \qquad = 615.04$$
$$\Sigma x^2 = 8607.69 \qquad \Sigma y^2 = 62.42$$

Substituting in the formula for the product moment correlation coefficient:

$$r = \frac{(10 \times 731.6) - (292.1 \times 24.8)}{\sqrt{[(10 \times 8607.69) - 85322.4][(10 \times 62.42) - 615.04]}}$$

$$r = \frac{(7316) - (7244.08)}{\sqrt{[754.45 \times 9.16]}} = \frac{71.92}{\sqrt{6910.762}} = \frac{71.92}{83.131}$$

$$r = 0.86$$

According to Table 13.1, there appears to be a strong positive correlation between chest circumference and birth weight in babies. We need to check that such a correlation is unlikely to have arisen by chance in a sample of 10 units (babies). Appendix 5 gives the probability distribution of r. First we need to work out the number of degrees of freedom. In the case of a correlation coefficient, they are the number of pairs of observations less 2, that is $(n - 2) = 8$ in our example. Consulting Appendix 5, we find that our calculated value of 0.86 exceeds the tabulated value at 8 df of 0.765 at $P = 0.01$. Our correlation is therefore statistically highly significant.

A measure of a baby's chest circumference is a good indication of its weight condition at birth, and could be adopted as a possible alternative to weighing in circumstances where accurate scales are not available.

13.5   The coefficient of determination r^2

The square of the product moment correlation coefficient is itself a useful statistic, and is called the **coefficient of determination**. It is a measure of the proportion of the variability in one variable that is accounted for by variability in another. In a perfect correlation where $r = +1$ or -1, a variation in one of the variables is exactly matched by a corresponding variation in the other. This situation is rare in the health sciences because many factors govern relationships between variables in patients. Thus, r^2 indicates to what extent other factors are influencing x and y. In Example 13.1 the coefficient of determination is $0.86^2 = 0.74$ (74%). It follows that some $100 - 74 = 26\%$ in the variation of weight is *not* accounted for by variation in chest circumference. The weight of the baby may be related to factors other than its chest circumference.

3.6 The Spearman Rank Correlation Coefficient r_s

In cases where the rather strict conditions of the Product Moment Correlation Coefficient do not apply, the Spearman Rank Correlation Coefficient may be used. This test is widely used by health scientists and uses *ranks* of the x and y observations and the raw data themselves are discarded.

The formula for calculating r_s is:

$$r_s = 1 - \left[\frac{6 \Sigma\, d^2}{(n^3 - n)} \right]$$

where n is the number of units (pairs of observations) in a sample, d is the difference between the ranks, Σ is the 'sum of' and 6 is a constant peculiar to this formula.

Because the observations are *ranks*, the test may be applied to observations measured on ordinal scales. It is suitable for samples with between 7 and about 30 pairs of observations. The following example shows how the formula is applied.

Example 13.2

Attendance at births by a trained health care worker varies enormously between countries. For example, in Bangladesh and Nepal it is reported that fewer than 10 per cent of births are attended; in Norway and Belgium, on the other hand, attendance is close to 100 per cent. Mortality of mothers during childbirth also varies. Could there be a correlation between mortality rate and attendance at the birth by a trained health care worker?

In Table 13.3, the x column shows a set of percentages of births attended for 12 different countries. For convenience, we have listed them in ascending order. The next column is the rank of each value of x (notice that there are two values of 82, and so the ranks have been averaged, as explained in Section 3.6). The y column is the maternal mortality rate per 100 000 live births, and the next column is the rank of each value. The column headed d is the difference between each pair of ranks, and d^2 is the square of these quantities.

Because 'percentage' is a derived variable, there may be doubt as to whether this variable is normally distributed, and so the Spearman Rank Correlation Coefficient is appropriate to the task.

Table 13.3   Attendance at births by trained care workers in different countries

Percentage of births attended (x)	Rank of x	Mortality rate of mothers (y) per 100,000 births	Rank of y	d	d^2
5	1	582	11	−10	100
24	2	450	10	−8	64
27	3	195	8	−5	25
29	4	284	9	−5	25
40	5	762	12	−7	49
57	6	90	6	0	0
70	7	25	4	3	9
82	8.5	61	5	3.5	12.25
82	8.5	124	7	1.5	2.25
87	10	12	3	7	49
96	11	11	2	9	81
99	12	5	1	11	121

$$\Sigma d^2 = 537.5$$

Substituting in the formula for r_s, we obtain:

$$r_s = 1 - \left[\frac{(6 \times 537.5)}{(12 \times 12 \times 12) - 12}\right] = 1 - \left(\frac{3225}{1716}\right) = (1 - 1.88) = -0.88$$

The value of r_s suggests a strong negative (inverse) correlation between percentage attendance and maternal mortality. However, we must first check that such a value could not have arisen by chance in a sample of 12 units if, in fact, there is no correlation within the population. Our statistical hypotheses are:

H_0: there is no correlation: the value of r_s is obtained by chance (sampling error).

H_1: a value of r_s as great as −0.88 could not be the result of sampling error in a sample size of 12 units.

The value of r_s is compared with the distribution shown in Appendix 4 (the negative sign is ignored). Entering the table at $n = 12$, we see that our calculated value of 0.88 exceeds the critical value of 0.780 at $P = 0.01$. We reject H_0 and say that the correlation is highly significant. We may infer that the absence of a trained health care worker present at a birth significantly increases the risk of death of the mother during childbirth.

Notice that this particular correlation is *negative*, or *inverse*. That is, the higher the percentage attendance, the lower is the mortality rate. You may wish to repeat this example and replace maternal *mortality rate* by *survival rate*. (Mortality rate is converted to survival rate by subtracting the mortality rate from 100 000: in the top row this is 100 000 − 582 = 99418.) The correlation coefficient is the same, but the value is positive, not negative.

3.7 Advice on measuring correlations

1. When observations of one or both variables are on an ordinal scale, or are proportions, percentages, indices or counts of things, use the Spearman Rank Correlation Coefficient. The number of units in the sample (the number of paired observations) should be between 7 and about 30. The ranking of over 30 observations is extremely tedious, and is not commensurate with any marginal increase in accuracy.

2. When observations are measured on interval scales, the use of the Product Moment correlation coefficient should be considered. Sample units must be obtained randomly, and the data should be *bivariate normal*, that is to say, both x and y observations should be normally distributed.

3. The relationship between the variables should be rectilinear (straight line) not curved. Certain mathematical transformations (e.g. logarithmic transformation) will 'straighten up' curved relationships.

4. Do not conclude that, because two variables are strongly and significantly correlated, one is necessarily the *cause* of the other. It is always possible that some additional, unidentified factor is the underlying source of variability in both variables.

5. Correlations measured in samples estimate correlations in the populations. A correlation in a sample is not 'improved' or strengthened by obtaining more observations; however, larger samples may be required to confirm the statistical significance of weak correlations.

14 REGRESSION ANALYSIS

4.1 Introduction

In Section 4.6 we illustrated the relationship between weight and chest circumference in babies by means of a scattergram. In presenting a scattergram, it is often helpful to draw a line through the cloud of points in such a way that the *average* relationship is depicted. The line is called the **line of best fit**. It has been added to the scattergram in Figure 14.1. A problem arises as to how to fit the best line through a cloud of points. If the scatter is not too great, the line may be reasonably fitted 'by eye'. In most cases, however, it is necessary to replace such a subjective method by a more objective mathematical approach. The line so produced is called a **regression line**. The regression line may be described in terms of a mathematical equation that defines the relationship between x and y, and we may use this equation to estimate or predict the value of one variable from a measurement of the other. We call this procedure (i.e. fitting the best line to a scattergram from an equation relating x and y)

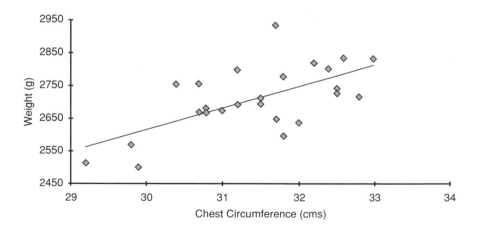

Figure 14.1   A regression line

regression analysis. Before we attempt to describe this important statistical technique, we must first consider a little basic geometry.

14.2   Gradients and triangles

The gradient of a hill slope is often expressed in such terms as 1 in 10. This means simply that for every 10 units of distance travelled in a horizontal plane an elevation of 1 unit in the vertical plane will result. The gradient, or slope, is symbolized by b and, in this case, is equal to $1 \div 10 = 0.1$. In the general case, the gradient is equal to an increment in y divided by a corresponding increment in x (Figure 14.2).

Knowing the value of b, we may use it to calculate the height gained from a given distance moved horizontally, thus:

Vertical height gained = gradient × horizontal distance travelled

In the general case, this may be expressed as:

$$y = bx$$

If we wish to know the actual final height rather than just the height gained, we must know the height of our starting point above some reference zero, say sea level. If this height is symbolized by a units, the final height is:

Final height = height of starting point + (gradient × horizontal distance travelled)

or, in the general case (see Fig. 14.3),

$$y = a + bx$$

The equation, $y = a + bx$ is known as the *equation of a straight line* or **rectilinear equation.** Regression analysis is concerned with solving the

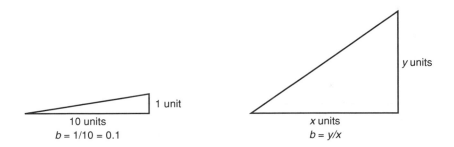

Figure 14.2   Gradients of slopes: gradient = y units / x units

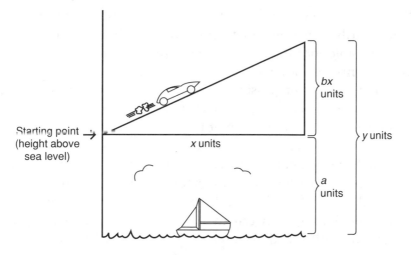

Figure 14.3 Final height of car above sea level $= a + bx$, where b is the gradient

values of a and b from a set of bivariate sample data. We may then accurately fit a line to a scattergram and estimate the value of one variable from a measurement of the other. The quantities a and b are both **regression coefficients**. In common usage, however, the term 're-gression coefficient' is taken to mean the slope of the regression line, b.

4.3   Dependent and independent variables

Up to this point we have not indicated which of two variables should be placed on the y (vertical) axis and which on the x (horizontal) axis. There is a convention that gives us a guideline in this respect. In many pairs of variables it is possible to discern, unambiguously, that one of the variables is 'dependent' on the other. For example, pulse rate might depend upon the amount of muscular exertion; the notion that the converse might be true, that the level of exertion in some way depends upon the pulse rate, is clearly ludicrous. In other examples we have given, the neonatal mortality rate of mothers appears to depend upon the presence of trained health care workers (Example 13.2); the height gained by the car in Section 14.2 depends upon the horizontal distance moved. The length that a spring is stretched depends upon the weight applied. The converse of these cannot be true. In these instances, the variable that describes the pulse rate, the maternal mortality, the height of the car and the length of the spring is the **dependent variable**. In each case, the other variable is the **independent variable**.

We should state here that in identifying a dependent variable we are not necessarily suggesting a *causal* relationship between the two. It is always possible that the two variables are independently related to a third, unidentified, variable.

Sometimes it is not possible to state which is the dependent and independent variable in a pair. Such is the case in the baby weight/chest circumference relationship of Example 13.1. If the regression line is to be used to estimate the value of one variable from a measurement of the other, then the variable that is used to estimate the other is placed on the *x*-axis; the variable *to be* estimated is placed on the *y*-axis. If, however, the regression line is calculated merely to describe the mathematical relationship between the two variables, and no dependent variable can be identified, then the choice of axis is arbitrary.

14.4   A perfect rectilinear relationship

By example we may relate the rectilinear equation derived in Section 14.2 to the idea of dependent and independent variables.

Example 14.1

There are many instances where sophisticated electronic scales and balances are not available, and spring balances are still used for health care purposes. An observer measures the length of a spring when known masses (weights) are suspended from it in order to calibrate it. The following data are recorded:

Mass x (g)	Length of spring y (cm)
10	10
20	15
30	20
40	25
50	30

Clearly, the length of the spring is dependent on the mass attached. We have no difficulty, therefore, in identifying mass as the *x* variable and length as the *y* variable. A scattergram drawn from these data is shown in

Figure 14.4(a). We see that all the points are in perfect alignment, and that the regression line may be drawn through them without the need for a mathematical computation (Figure 14.4(b)). When extrapolated downwards, the line cuts the y-axis at point a. This is the **intercept on the y-axis**. It represents the length of the spring in its unstretched state, and is analogous to the *height above sea level* in the example in Section 14.2. We see from the scale on the y-axis that a has a value of 5 cm. Thus, one of the quantities in the $y = a + bx$ equation has been determined.

To calculate the other, b, drop a vertical line from any point on the regression line and complete a right-angled triangle with a horizontal line, as shown in Figure 14.4(c). The gradient b is the number of units on the y-scale divided by the number of units on the x-scale that correspond to the respective sides of the triangle. In Figure 14.4(c), $b = 15 \div 30 = 0.5$.

We now have both quantities in the equation $y = a + bx$. Thus:

$$y = 5 + 0.5x$$

Using this equation, we may estimate the length of the spring when a mass of any size is attached. We can check this for a value we already know. An observation in the table above shows that a mass of 50 g produces a length of 30 cm. Does the equation bear this out?

$$y = 5 + (0.5 \times 50) = 5 + 25$$
$$y = 30 \, \text{cm}$$

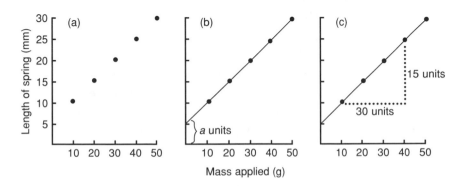

Figure 14.4   A perfect linear relationship

Clearly, we have obtained the correct result for the values of *a* and *b*. We may confidently interpolate between the points and predict the length for, say, a mass of 25 g.

$$y = 5 + (0.5 \times 25) = 17.5 \, \text{cm}$$

In this way, we can build up a series of *y* values to produce a finely graded scale. If the scale were to be etched onto a stick aligned vertically beside the spring, we would have a rudimentary spring balance. Although it is safe to interpolate between points on the scattergram, caution should be exercised in extrapolating far beyond the last point. If a 500 g mass is hung from the end of our spring, it might well overload the capacity of the spring to return to its original length, and it is by no means certain that its stretched length is equal to that predicted by the equation.

As we have noted already, such perfect rectilinear relationships may be encountered in physics, but seldom in health or life sciences, where a cloud of points rather than a nice straight line is more usual.

14.5   The line of least squares

Fitting a regression line to a scattergram involves placing it through the points so that the sum of the vertical distances (**deviations**) of all points from the line is minimized. Because some deviations are negative and some positive, it is more convenient to utilize the sum of the *squares* of the deviations, Σd^2. In this way, awkward negative signs are avoided. The method of fitting the line is therefore known as the **method of least squares**. Figure 14.5 shows the line of least squares fitted to four points in which the squares of the vertical deviations are minimized. Any alternative position of the line (up or down, or with different slope) will increase Σd^2. Although it is possible to reduce the values of, say, $d^2{}_1$ and $d^2{}_4$ by moving the line closer to these points, it is only at the expense of increasing $d^2{}_2$ and $d^2{}_3$. Since we are dealing with *squares*, the increase in $(d^2{}_2 + d^2{}_3)$ is not offset by the decrease in $(d^2{}_1 + d^2{}_4)$.

Fitting the line in this way, where Σd^2 for vertical deviations (deviations on the y-axis) is minimized, is called the **regression of y on x**, and is usually referred to as **simple linear regression**.

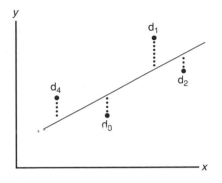

Figure 14.5 Deviations from the regression line

4.6   Simple linear regression

Regrettably, the above title belies some rather strict conditions that apply to the use of least squares regression. The main conditions are:

1. There is a linear relationship between a dependent y variable and an independent x variable that is implied to be *functional* or *causal*.

2. The x variable is not a random variable but is under the control of the observer. Regression is therefore especially applicable to *experimental* situations in which, for example, the response of a y variable to predetermined quantities of time, temperature, amount of drug, etc. is under investigation.

3. The scatter of points on either side of the regression line are about the same over the whole length of the line: they should not be close to the line at one end, and 'fan out' towards the other.

There are many examples of regression in the research literature where condition (2) does not strictly hold true. However, so long as the *dependent* variable, and in particular, the variable of which predictions are *to be* made, is allocated to the y axis, then for practical purposes, statisticians regard the error arising as acceptably small.

Example 14.2

Investigators at a sports health centre are interested in the relationship between oxygen consumption and exercise time in athletes recovering from injury. Appropriate mechanisms for exercising and measuring oxygen consumption are set up (the practical details are not important in this example), and the results are presented below.

x variable Exercise time (min)	0.5	1.0	1.5	2.0	2.5	3.0	3.5	4.0	4.5	5.0
y variable Oxygen consumption	620	630	800	840	840	870	1010	940	950	1130

A scattergram of these data (Figure 14.6) reveals a roughly rectilinear relationship; within the limits of the study, the oxygen consumption clearly depends upon the duration of exercise, and a functional relationship is presumed. Observations on the x-axis are under the control of the observer, and so simple linear regression is appropriate. Notice that the x observations are not normally distributed – they are evenly spread over a range of $0.5 - 5.0$ min, rather than being clustered around a mean value.

The information we require to calculate a and b is the same as that needed to calculate the product moment correlation coefficient. In addition, we must obtain the mean of the x observations ($\bar{x}$) and the mean if the $y(\bar{y})$ observations. Using a calculator, we determine:

$$\bar{x} = 2.75 \qquad\qquad \bar{y} = 863 \qquad\qquad n = 10$$
$$\Sigma x = 27.5 \qquad\qquad \Sigma y = 8630$$
$$(\Sigma x)^2 = 756.25 \quad (\Sigma y)^2 = 74\,476\,900 \quad \Sigma xy = 25\,750$$
$$\Sigma x^2 = 96.25 \qquad \Sigma y^2 = 7\,672\,500$$

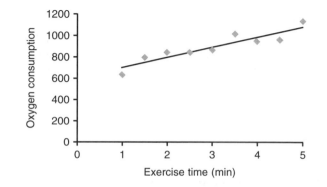

Figure 14.6   Regression of oxygen consumption on exercise time

The computation of a and b for the regression of y on x requires two steps. First, the calculation of b, and then, using the value of b so derived, the calculation of a. The formula for the calculation of b is:

$$b = \frac{n\Sigma xy - \Sigma x \Sigma y}{n\Sigma x^2 - (\Sigma x)^2}$$

Substituting in the formula:

$$b = \frac{(10 \times 25750) - (27.5 \times 8630)}{(10 \times 96.25) - (756.25)} = \frac{20175}{206.25}$$

$$b = 97.82$$

To calculate a, we use the rectilinear equation, $y = a + bx$ or, rearranging, $a = y - bx$. We have derived a value for b, but which of the 10 possible values of x and y should we use to solve the equation for a? Since all points constitute a *scatter*, it is most unlikely that any one single pair will be correct. The answer is to use the mean of x and the mean of y thus:

$$a = \bar{y} - b\bar{x}$$

We now have all the terms needed to solve for a:

$$a = 863 - (97.82 \times 2.75)$$
$$a = 863 - 269$$
$$a = 594$$

The regression equation may now be written out in full:

$$y = 594 + 97.82x$$

We can use the regression equation to predict the amount of oxygen consumed from any given time of exercise. Thus, for an exercise time of 2.8 min we predict:

Oxygen consumption $= 594 + (97.82 \times 2.8) = 868$ units

It is only safe to make predictions within the range of the initial x observations. We cannot be sure that the regression line will continue to be linear after, say, several hours of exercise!

14.7   Fitting the regression line to the scattergram

To fit a line to a scattergram we need to know the positions of two points on the line. The further away apart the points are, the more accurately can the line be drawn. The points are calculated from the regression equation, $y = a + bx$. Two separate values of x are selected to derive corresponding y values from the equation. These become the coordinates for the two points.

Select a value of x that is close to the y axis. In Example 14.2 this could be 0.5 min. Calculate the corresponding y value from the equation:

$$y = 594 + (97.82 \times 0.5) = 642.9$$

Mark a point on the scattergram at coordinates $x = 0.5$; $y = 642.9$. If the scale is scaled to start at zero, then the y coordinate at $x = 0$ is of course equal to a, the intercept. The point can be marked here on the y-axis. Select a second value of x, far from the y-axis, say, 5.0 min. As before, calculate the corresponding y value from the equation:

$$y = 594 + (97.82 \times 5.0) = 1083.1$$

Mark a second point on the scattergram at coordinates $x = 5.0$; $y = 1083.1$. Join the two points with a straight line, but do not exceed the cloud of points by very much at each end.

14.8   Regression for estimation

In Section 13.4, Example 13.1, we described a situation in which a health care worker was attempting to find an alternative variable to weight in order to assess the 'condition' of new-born babies, when full clinical facilities were not available. We establish that there is a strong correlation between weight and chest circumference. Therefore, an assessment of birth condition can be made with as simple a facility as a tape measure.

If the relationship between weight and chest circumference is a good one, there is the possibility that the relationship could be defined in terms of a regression equation that could be used, in principle, to estimate the weight of a baby from a measurement of its chest circumference. All the data needed to do this is already presented in Section 13.4.

However, we must note two important points before proceeding. First, there is no clear dependent variable. It is no more true to say that weight depends upon chest circumference than vice versa. Secondly, the condition relating to 'observations on the x-axis being under the control of the observer' (condition '2' in Section 14.6) is clearly not true.

Despite these limitations, statisticians approve the use of simple linear regression under these circumstances, provided that (a) the variable *to be estimated* (weight in this case) is assigned to the y-axis, so that it is effectively the dependent variable; and (b), rather than select sampling units (babies in this case) randomly, units are selected to span the range of measurements available. Treating the data presented in Example 13.1 in the way that we did in Example 14.2, we calculate that $a = -0.304$ and that $b = 0.095$. Therefore:

$$\text{Weight} = -0.304 + (0.095 \times \text{chest circumference})$$

Therefore, a baby whose chest circumference is 31.0 cm has an estimated weight of 2.64 kg. In the absence of accurate scales, this estimation might prove to be useful under some circumstances.

14.9   The coefficient of determination in regression

In Section 13.5 we introduced the useful term **coefficient of determination**, which is equal to the square of the correlation coefficient. In regression analysis, because sampling units are rarely drawn *randomly* from the population (remember, the observations on the x axis are supposed to be chosen by the observer, not selected randomly), the correlation coefficient has no meaning. However, its square, the coefficient of determination, does have a useful meaning. As before, it is a measure of the proportion of the variability in one variable that is accounted for by variability in the other. For the oxygen consumption exercise case described in Example 14.2, we calculate r to be 0.937, and thus $r^2 = 0.878$. In other words, 87.8 percent of the variability in oxygen consumption is accounted for by variability in duration of exercise. Values of r^2 which exceed about 70 percent indicate extremely strong relationships between the variables.

14.10   Dealing with curved relationships

We noted in Section 14.6 that one of the conditions applicable to simple linear regression is that there is a *linear* relationship between the x and y variables. Many strong relationships in health science are not linear, but exhibit *curved* lines of best fit. Growth and mortality are examples of this. The usefulness of regression is greatly increased when curved relationships are 'straightened up' by *transformation*. It then becomes possible to undertake regression analysis, and to estimate a value of one variable from a measurement of the other.

Example 14.3

A patient is given an intravenous bolus dose of 10 mg of a drug (we call this 'Drug D'), and plasma concentrations of the drug are measured at timed intervals afterwards. The data are presented in the table below. Included in the table are the $\log_{10}$ values of the y variable.

Time (min) (x)	Plasma concentration of Drug D (μ g/L)(y)	$\log_{10}$ concentration of Drug D (y')
10	1100	3.04
20	800	2.90
30	750	2.88
40	630	2.80
50	540	2.73
60	530	2.72
70	480	2.68
90	380	2.58
110	350	2.54
150	240	2.38

The relationship between plasma concentration of Drug D and time is shown in Figure 14.7(a); it is distinctly curved. However, when the drug concentrations are transformed to their logarithms (y') and plotted against time, the curve is obviously 'straightened up', as shown in Figure 14.7(b). The transformation appears to be satisfactory.

Using a calculator to process and summarize the data (note that, for simplicity, we denote log y as y'):

$$\Sigma x = 630 \qquad \Sigma y' = 27.25 \qquad n = 10$$
$$(\Sigma x)^2 = 396\,900 \quad (\Sigma y')^2 = 742.56 \quad \Sigma xy' = 1642.7$$
$$\Sigma x^2 = 56\,700 \qquad \Sigma y'^2 = 74.59$$
$$\bar{x} = 63 \qquad \bar{y}' = 2.73$$

Substituting in the formula for b (Section 14.6):

$$b = \frac{[(10 \times 1642.7) - (630 \times 27.25)]}{(10 \times 56700) - 396900} = \frac{-740.5}{170100} = -0.00435$$

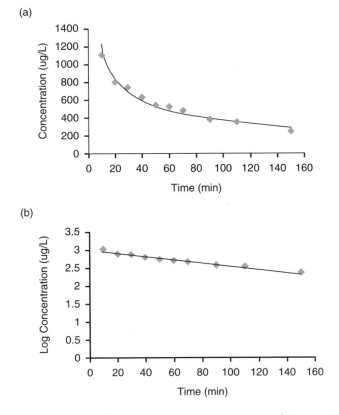

Figure 14.7 Graph of decline in plasma concentration of 'Drug D'. (a) Axes untransformed; (b) y-axis transformed logarithmically

The negative value of b indicates the declining slope of the graph. Solving for a as before, but remembering that this time a is measured on a logarithmic scale, we designate it a':

$$a' = \bar{y}' - b\bar{x} = 2.73 - (-0.00435 \times 63)$$

(*Hint: remember that the double negative represents an addition*)

$$a' = 3.00$$

We may now substitute the coefficients into the regression equation:

$$y' = a' + bx$$

That is,

$$y' = 3.00 + (-0.00435 \times x)$$

To test the veracity of the equation, we can use it to predict the plasma concentration of Drug D after 55 minutes ($x = 55$):

$$y' = 3.00 + (-0.00435 \times 55) = 2.761$$

To back transform y' to y, take the antilog:

$$\text{Antilog } 2.761 = 577\,\mu g/L$$

For readers with an interest in mathematics, there is an alternative way of writing the logarithmic regression equation, namely:

$$\text{Log } y = \log a + bx$$

Or, in exponential form (by taking antilogs on both sides):

$$y = a \times 10^{bx}, \text{ where } a \text{ is antilog } a' = 1000$$

Thus, the plasma concentration of Drug D after 55 min is estimated to be:

$$y = 1000 \times 10^{(-0.00435 \times 55)}$$
$$= 1000 \times 0.5764 = 576.4\, \mu g/L$$

4.11 How can we 'straighten up' curved relationships?

In Example 14.3 we 'straightened up' the relationship by transforming the observations on the y-axis to their logarithms, and by plotting log y against x. However, this is not always the correct solution. In other cases, transformation of values on the x-axis, or transformation of values on *both* axes might have resulted in a better straightening effect. How should we know what to do?

First, there is often a well-established underlying mathematical relationship between variables (this is sometimes called a 'model'). Thus, the logarithmic decline in plasma concentration of a drug illustrated in the preceding example is well known. Readers with knowledge of biochemistry may be aware of the classical regression known as the Lineweaver – Burke plot, in which the curved relationship between velocity of an enzyme-catalyzed reaction and substrate concentration is straightened up by a *reciprocal* transformation of both axes; that is, x is replaced by 1/x and y is replaced by 1/y.

Alternatively, if you are not aware of an underlying model, you may simply employ 'trial and error'. First transform one axis, then the other, and finally both, to see which alternative give the best straightening effect. The value of the *coefficient of determination* (Sections 13.5 and 14.9) may assist you in deciding on the most effective transformation, the higher value being the one to choose. For example, the coefficient of determination of the *untransformed* data in Example 14.3 is 82.4 percent; after transformation it increases to 96.0 percent. Clearly, the transformation is effective.

4.12 Advice on using regression analysis

1. A regression equation can be derived for any set of bivariate data simply by substituting them in the formulae. Because we *can* undertake a particular mathematical treatment, however, it does not mean that we *should*. The use of regression analysis should be restricted to cases where it is necessary to place a *best line* though a cloud of

points and, in particular, where the estimation of one variable from a measurement of the other is required. It is a common fallacy to imagine that because an equation has been solved, the quality of the analysis is somehow enhanced. If all that is required is a measure of the strength of the relationship between two variables, the use of the correlation coefficient may be more appropriate.

2. Before proceeding with the mathematics of regression, always draw a scattergram of the data to see what it *looks* like. From this you can decide if two important conditions are met: (i) are the points roughly linear, not curved? (ii) is the scatter of points around the line reasonably even over the whole length of the line.

3. If the scattergram suggests a *curved* relationship, then transformation of one or other or both axes will usually straighten up the line. We have given an example of a logarithmic transformation, but arcsine transformation for proportions (Section 9.11) or square root transformation (for counts of things) may prove effective.

4. The correlation coefficient r has no meaning in regression unless sample units have been obtained randomly and observations are normally distributed on both axes. However, the *coefficient of determination* r^2 is a useful number. It tells us the proportion of the variability in the y observations that can be accounted for by variability in the x observations.

15 COMPARING AVERAGES

15.1 Introduction

Health care scientists often wish to compare the average value of some variable in two samples. The problem typically resolves into two steps:

1. Is the observed difference between the average of two samples *significant*, or is it due to chance sampling error that we described in Section 10.1?

2. If a difference between the average of two samples is indeed *significant*, what is the extent of the difference?

We have already dealt with the second step in Sections 10.4 and 10.5. In this chapter, we outline methods that test the significance of a difference between the average of two samples. In each method we describe, the null hypothesis is similar:

H_0: samples are drawn from populations with identical averages and any observed difference between the samples is due to sampling error.

The alternative hypotheses are:

H_1 (two-tailed test): samples are drawn from populations with different averages; an observed difference between samples cannot be accounted for by sampling error; or

H_1 (one-tailed test): a nominated sample is drawn from a population that has a larger average than that from which the other is drawn. (As we have warned in Section 11.5, we recommend that you consider all tests to be *two-tailed* unless you have strong grounds for considering otherwise.)

Tests for difference between sample averages may be parametric or non-parametric. Non-parametric tests convert observations to ranks and compare sample distributions, essentially their *medians*. Parametric tests use actual observations and compare *means*. Parametric tests are considered to be more 'powerful' than non-parametric only if the conditions applying to their use are fulfilled (Section 11.8). We describe non-parametric methods first, and then proceed to parametric methods.

15.2   Matched and unmatched observations

When analysing bivariate data such as correlations, a single sample unit gives a pair of observations representing two different variables. The observations comprising a pair are uniquely linked, and are said to be **matched**. It is also possible to obtain a pair of observations of a *single* variable that are matched. For example, the systolic blood pressure measurements of 10 patients and the measurements of another 10 after medication for hypertension are unmatched. However, the measurements of the *same* 10 patients before and after the adminis-tration of the medication are matched, because the measurements of 'before and after' can be uniquely paired off according to patient. It is possible to conduct a more sensitive analysis if the observations are matched.

15.3   The Mann–Whitney *U*-test for unmatched samples

The Mann–Whitney *U*-test is a non-parametric technique for comparing the medians of two unmatched samples. Because the values of the observations are converted into their *ranks*, the test may be applied to variables measured on ordinal or interval scales. Moreover, because the test is 'distribution-free' it is suitable for data which are not normally distributed, for example, counts of things, proportions or other derived variables. Sample sizes may be unequal. The task is to compute a test statistic U that is compared with tabulated critical values (Appendix 6).

Example 15.1

The number of patients with bone fractures attending a city's Accident and Emergency departments on Saturdays and Sundays over eight weekends in winter are presented below. To emphasize the point that sample sizes do not have to be equal, we have assumed that the record from one department is not available for the final Sunday, and is therefore missing. We wish to know if the apparently higher number of accidents on Saturdays is due to chance variation, of if there is an underlying cause.

It is always worthwhile to remind yourself of the definition of the populations being sampled. Since we are dealing with *counts* (numbers of patients), the sampling unit is a Saturday (or a Sunday). The sample is the number of Saturdays (or Sundays) for which we have an observation, and the population is a hypothetical population of Saturdays (or Sundays). The null hypothesis H_0 is therefore that the two samples are from populations with equal averages (medians).

In each sample the observations are tabulated in ascending order for convenience:

Sample 1								
Saturdays:	8	12	15	21	25	44	44	60
Sample 2								
Sundays:	2	4	5	9	12	17	19	

The median in Sample 1 is 23, considerably larger than the median of 9 in Sample 2 (see Section 7.4). (Note: it is not necessary to calculate the median in order to perform the test.) However, there is considerable overlap between the observations in the samples. A test is required to decide if the difference is statistically significant. The Mann–Whitney *U*-test is appropriate. The procedure for using the test is as follows:

1. List all observations in both samples in ascending order, assigning them ranks. Where there are tied ranks, the average rank is assigned, as explained in Section 3.6. Distinguish between the samples by underlining one of them. We have underlined the observations in Sample 1.

Observation:	2	4	5	8	9	12	12	15	17	19	21	25	44	44	60
Rank:	1	2	3	4	5	6½	6½	8	9	10	11	12	13½	13½	15

2. Sum the ranks of each sample; that is, sum the ranks of the under-lined and non-underlined separately. Let R_1 = sum of the ranks of Sample 1 and R_2 = the sum of the ranks of Sample 2:

$$R_1 = 4 + 6\tfrac{1}{2} + 8 + 11 + 12 + 13\tfrac{1}{2} + 13\tfrac{1}{2} + 15 = 83.5$$

$$R_2 = 1 + 2 + 3 + 5 + 6\tfrac{1}{2} + 9 + 10 = 36.5$$

3. Calculate the test statistics U_1 and U_2 from

$$U_1 = n_1 n_2 + \frac{n_2(n_2 + 1)}{2} - R_2 = 56 + 28 - 36.5 = 47.5$$

$$U_2 = n_1 n_2 + \frac{n_1(n_1 + 1)}{2} - R_1 = 56 + 36 - 83.5 = 8.5$$

4. To confirm that our ranks have been correctly assigned and summed, check at this point that:

$$U_1 + U_2 = n_1 n_2$$

$$47.5 + 8.5 = 56 = 8 \times 7$$

This being the case, proceed to Step 5.

5. Select the *smaller* of the two U values (i.e. $U_2 = 8.5$ in this example), and compare it with the value in the table (Appendix 6) for the appropriate values of n_1 and n_2 (8 and 7 in this example). From the table, the critical value (at $n_1 = 8$ and $n_2 = 7$) is 10. Our smaller value of U is *less* than the critical value. The null hypothesis is therefore rejected. There is a statistically significant difference be-tween the medians ($U = 8.5$; $P < 0.05$, Mann–Whitney U-test).

15.4   Advice on using the Mann–Whitney U-test

1. The Mann–Whitney U-test may be applied to interval data (meas-urements), to counts of things, derived variables (proportions and indices) and to ordinal data (rank scales, etc.). It may be used with as few sampling units in each sample as 2 and 8, 5 and 3, or 4 and 4. However, with samples as small as these, there must be no overlap of observations between the two samples for H_0 to be rejected. Since

this can be determined by inspection of the data, it is hardly worth proceeding with the test.

2. Note that, unlike some test statistics, the calculated value of U has to be *smaller* than the tabulated critical value in order to reject H_0.

3. The test is for a difference in *medians*. It is a common error to record a statement like 'the Mann–Whitney U-test showed there is a significant difference in means'. There is, however, no need to calculate the medians of each sample to do the test.

4. Although there is no requirement for the observations in the samples to be normally distributed, the test does assume that the two distributions are similar. It is not permissible, therefore, to compare the median of a positively skewed distribution with that of a negatively skewed one. Since it is usually impossible to identify the shape of a frequency distribution in small samples, the point is largely academic. Nevertheless, if it is known from other studies that the two samples have been drawn from populations that have fundamentally different shaped frequency distributions, then the test should not be used.

15.5 More than two samples – the Kruskal–Wallis test

The Mann–Whitney U-test is designed to compare the averages of two samples. Sometimes, however, we wish to compare the averages of several samples. Suppose a hospital administrator wishes to compare average waiting times in four different outpatient clinics, A, B, C and D. Five patients are sampled randomly from each clinic, thus generating four samples of five observations. It is possible to compare the medians using the Mann–Whitney U-test, but the test would have to be repeated six times to compare A with B, A with C, and so on for all combinations. Apart from being extremely tedious, there is an important statistical reason for avoiding *multiple comparisons* of this type. We explain the reason fully in Section 16.1. The Kruskal–Wallis test is a simple non-parametric test to compare the medians of three or more samples. Observations may be interval scale measurements, counts of things, derived variables or ordinal ranks. If there are only three samples, then there must be at least five observations in each sample. Samples do not have to be of equal sizes. The method of performing the test is explained in the following example.

Example 15.2

A health worker collects the waiting times of five patients selected randomly from four different out-patient clinics, A, B, C and D. Are there differences between the average waiting time at each clinic? The steps in performing the Kruskal–Wallis test are as follows and relate to Table 15.1.

1. Tabulate the observations in columns for each sample A, B, C and D. Assign to each observation its *rank within the table as a whole*. If there are tied ranks, assign to each its average rank as described in Section 3.6. Place each rank in brackets beside its observation.

2. Write the number of observations n in each sample (five in each in this example) in a line under its respective column. Add these up to obtain N, the *total* number of observations (20 in this case).

3. Sum the ranks of the observations in each sample (i.e. the numbers in brackets) and write them (R) in the next line under n.

4. Square the sum of ranks and write these (R^2) in a line under R.

5. Divide each value of R^2 by its respective value of n. Write this (R^2/n) in the bottom line under R^2. Add up the separate values of R^2/n to obtain $\Sigma(R^2/n)$.

Table 15.1   Waiting times (min) at four outpatient clinics (rank score in brackets)

	A	B	C	D	
	27 (12)	48 (16)	11 (6)	44 (15)	
	14 (7)	18 (9½)	0 (1)	72 (19)	
	8 (4½)	32 (13)	3 (2)	81 (20)	
	18 (9½)	51 (17)	15 (8)	55 (18)	
	7 (3)	22 (11)	8 (4½)	39 (14)	
n	5	5	5	5	$N = 20$
R	36	66.5	21.5	86	
R^2	1296	4422.25	462.25	7396	
R^2/n	259.2	884.45	92.45	1479.2	$\Sigma(R^2/n) = 2715.3$

The results of steps 1 to 5 are shown in Table 15.1.

6. The test statistic, K is obtained by multiplying $\Sigma(R^2/n)$ by a factor $\frac{12}{N(N+1)}$ and then subtracting $3(N+1)$ where the numbers 12 and 3 are constants peculiar to the formula.

$$K = \left[\Sigma(R^2/n) \times \frac{12}{N(N+1)}\right] - 3(N+1)$$

$$K = \left[2715.3 \times \frac{12}{20(21)}\right] - 3(21)$$

$$K = 14.58$$

7. Compare K with the tabulated distribution of χ^2 (Appendix 3). The degrees of freedom are the number of samples less one ($4 - 1 = 3$ in this example). At 3 df our calculated value of 14.58 exceeds the tabulated value of 11.34 at $P = 0.01$. We reject the null hypothesis, and conclude that there is a highly significant difference between the average patients waiting time between the four clinics.

It should be remembered that the test is applied to the samples *as a group*, and that we are confident that there are differences within the group as a whole. We should be cautious in making inferences about differences between particular pairs of samples, or between one sample and the remaining ones. However, it is safe to assume that *at least* there is a significant difference between the two samples which have the highest and the lowest sum of ranks. Inspecting the table of data, we note that these are sample D and sample C, respectively. We infer that these two, at least, are significantly different from each other.

15.6 Advice on using the Kruskal–Wallis test

1. Apply the test to compare the locations (averages) of three or more samples. If there are only three samples there should be more than five observations in each sample.

2. The test statistic, K, is compared with the distribution of χ^2. This does not, however, mean that observations have to be frequencies.

Data may be on any scale of measurement that allows ranking and need not be normally distributed.

3. If the outcome of the test suggests rejection of the null hypothesis, be cautious about making *a posteriori*, that is, unplanned comparisons between the samples, other than between the two with the highest and lowest sum of ranks. Having said that, 'common sense' should be used. It may be perfectly obvious from an inspection of the distribution of ranks that, for example, a particular sample stands out from the remainder.

15.7 The Wilcoxon test for matched pairs

The Wilcoxon test for matched pairs is a simple non-parametric test for comparing the medians of two matched samples. It calls for the calculation of a test statistic T whose probability distribution is known. In the test, one observation in a matched pair is subtracted from the other. Observations must therefore be measured on an *interval* scale. It is not possible to use this test on ordinal measurements.

Example 15.3

An investigator wishes to conduct a simple trial to demonstrate the effect of alcohol consumption on the performance of a manipulative task, using a group of students and a supply of wine. To test the significance one could use the Mann–Whitney U-test to compare the median time undertaken to perform the task by a sample of students who had not drunk wine with that of another sample who had drunk a standardized volume of wine. However, if the *same* students were timed before and after the consumption of wine, the two times from each student would constitute a *matched pair*. As we noted in Section 15.2, it is possible to conduct a more sensitive analysis with matched data than with unmatched data.

Fifteen students were timed doing the same complex task before and after imbibing a quantity of wine. The results are shown in Table 15.2. Did the wine affect their performance?

Scanning Table 15.2, it appears that most, but not all, students took longer to undertake the task after the consumption of wine. To what

Table 15.2 Time taken by 15 student subjects (A–O) to undertake a task before and after the consumption of wine

Subject	Pre-wine Sample A (min)	Post-wine Sample B (min)
A	15.1	21.2
B	14.5	17.8
C	16.5	17.8
D	19.2	18.8
E	16.9	19.2
F	14.3	17.1
G	16.4	16.2
H	22.5	27.0
I	20.4	22.8
J	19.5	18.5
K	14.2	21.8
L	16.9	16.9
M	20.5	21.7
N	26.2	37.5
O	12.2	14.5

extent is this result statistically significant, or could the observed outcome be the result of chance? The Wilcoxon test is appropriate to the problem. The null hypothesis H_0 is that there is no difference between the median times of the two sets of data. The alternative hypothesis H_1 is that there *is* a difference, but with no prediction which way the difference will lie (i.e. a two-tailed test).

If H_0 is true, we have two expectations, namely:

1. The number of time *increases* will be matched by a similar number of time *decreases*. If H_1 is true there will be more of one that the other.

2. The *sizes* of any time differences will be balanced evenly between increases and decreases; if H_1 is true, there will be a tendency for the larger differences to be in one direction.

The Wilcoxon test for matched pairs quantifies both the direction and magnitude of all the changes in a set of matched pairs. The steps to carry out the test are as follows:

1. For each matched pair, subtract the value of observation A from the value of observation B. The answer is the difference, *d*. The value of *d* has a negative sign if A is larger than B. It does not matter which sample is called A or B.

2. Rank the values of d according to their absolute values. That is to say, ignore the plus and minus signs for the moment. Ignore any instances in which $d = 0$ (there is one such case in this example). If any ranks are tied, assign average ranks exactly as described in Section 3.6.

3. Assign to each rank a '+' or a '−' sign corresponding to the sign of d.

4. Sum the *ranks* of the plus values and the minus values of d separately. The results of steps 1–4 are given in Table 15.3:
   Sum of the minus ranks: $2 + 1 + 3 = 6$
   Sum of the plus ranks: $12 + 10 + 5 + 6½ + 9 + 11 + 8 + 13 + 4 + 14 + 6½ = 99$
   The *smaller* value of the two sums of ranks is the test statistic T. In this Example, $T = 6$.

5. Consult the table of the probability distribution of T (Appendix 7). When T is *equal to or less than* the critical value in the table, the null hypothesis is rejected at the particular level of significance. Enter the table at the appropriate value of N. N is not necessarily the total number of pairs of data, but the number of pairs less the number of pairs for which $d = 0$. There is one in the present example (student L), and so $N = 14$. Entering the Appendix at $N = 14$, we find that our calculated value of T is *less* than the tabulated value of 21 under

Table 15.3   Ranking of matched pairs

Subject	Sample A	Sample B	d	Rank of d
A	15.1	21.2	+6.1	+12
B	14.5	17.8	+3.3	+10
C	16.5	17.8	+1.3	+5
D	19.2	18.8	−0.4	−2
E	16.9	19.2	+2.3	+6½
F	14.3	17.1	+2.8	+9
G	16.4	16.2	−0.2	−1
H	22.5	27.0	+4.5	+11
I	20.4	22.8	+2.4	+8
J	19.5	18.5	−1.0	−3
K	14.2	21.8	+7.6	+13
L	16.9	16.9	0	—
M	20.5	21.7	+1.2	+4
N	26.2	37.5	+11.3	+14
O	12.2	14.5	+2.3	+6½

the $P = 0.05$ column. We therefore reject H_0, and conclude that the 'pre' and 'post' wine values are statistically significantly different.

6. Record the result of the test as: 'there is a statistically significant difference between the median times of the two sample ($T = 6$, $P < 0.05$, Wilcoxon test for matched pairs'.

We emphasize that this rather crude experiment is just for illustrative purposes. As a result of this simple trial, the investigator is justified in pursuing a hunch that alcohol consumption affected performance by designing a more sophisticated trial. Ideally, some 'blinding' is involved, in which the subjects do not know if they are drinking alcohol or not. See Chapter 5 for more information about clinical trials.

There are other ways in which pairs of data may be matched. For example, to eliminate a diversity of genetically controlled variables, pairs of identical twins could be used for 'with and without' trials. The observations from each pair of twins then comprise a natural 'matched pair'.

5.8   Advice on using the Wilcoxon test for matched pairs

The Wilcoxon test for matched pairs may only be applied when the value of one observation in a matched pair can be subtracted from the other. That is, they should be interval measurements or counts of things. The number of matched pairs whose difference is not zero should be six or more. If the number of matched pairs exceeds about 40, the test is cumbersome and a parametric alternative (Section 15.12) is more appropriate.

1. Note that for H_0 to be rejected, T has to be *smaller* than or *equal to* the tabulated value at a given probability value.

2. The test is for a difference in *medians*. Do not be tempted to make statements about sample *means*.

3. The test assumes that samples have been drawn from parent populations that are symmetrically but not necessarily normally distributed. Because it may be impossible to discern the shape of the distribution in small samples, the point is largely academic. Nevertheless, if it is known from other studies that the two samples have been drawn

from populations that have fundamentally asymmetrical distributions, do not use the test.

15.9   Comparing means – parametric tests

Parametric tests that compare means are more restrictive than their non-parametric counterparts. First, data should be recorded on interval or ratio scales of measurement. Secondly, data should be approximately normally distributed; that they are so can be checked as described in Section 9.10. It is possible to transform to normal certain data, for example counts of things and proportions, by appropriate transformation (see Section 9.11). A third restriction is that the populations from which the samples are drawn should have similar variances (that is, standard deviation squared, Section 8.6). There are tests available to check that this is so, but so long as the larger of the two sample variances is not more than about four times that of the smaller, it is considered save to proceed with the test. The null hypothesis in a test for a difference between sample means is therefore:

H_0: two samples are drawn from populations with identical means and variances.

If it assumed that the 'parent' populations have similar variances, then a 'significant' outcome of a test may be attributed to a difference between population *means*.

15.10   The z-test for comparing the means of two large samples

When two populations have identical means, then $\mu_1 = \mu_2$ and the population mean difference $(\mu_1 - \mu_2) = 0$. The value of a difference between the means of two *samples* drawn from these populations can be expressed as $(\bar{x}_1 - \bar{x}_2)$. If H_0 is true, we would expect this difference to be close to zero, with any difference form zero being attributed to sampling error. The z-test is simpler than the next test we describe (the t-test), but requires larger sample sizes. Sample sizes exceeding 30 are considered adequate. The formula for the z-test requires the input of the means of the two samples, the standard deviations and number of observations. The test statistic which is calculated is z which we know (Section 9.7) has to exceed 1.96 to reject H_0 at the $P = 0.05$ level, or 2.58 at the $P = 0.01$ level. The formula is:

$$z = \frac{(\bar{x}_1 - \bar{x}_2)}{\sqrt{\dfrac{s_1^2}{n_1} + \dfrac{s_2^2}{n_2}}}$$

It does not matter which sample is nominated sample 1, but if $\bar{x}_2$ is larger than $\bar{x}_2$, then z will have a negative sign. This is ignored in two-tailed tests (Section 11.5).

Example 15.4

A health investigator suspects that pollution in cities reduces lung function, as compared to those living in a rural environment. The lung function (as measured by morning peak expiratory flow rate, millilitres per min.) of a sample of patients located at a city hospital were compared with a second sample from hospital with a predominantly rural catchment population. The basic sample statistics are given below:

Town (Sample 1): $\bar{x}_1 = 1600$; $s_1 = 41.0$; $n_1 = 150$
Country (Sample 2): $\bar{x}_2 = 1660$; $s_2 = 55.0$; $n_2 = 90$

H_0 = the two samples are drawn from populations with equal means;
H_1 = the two sample are drawn from populations with different means.
It is assumed that the variances of the two populations are similar.
  Substituting in the formula for z:

$$z = \frac{(1600 - 1660)}{\sqrt{\left(\dfrac{1681}{150}\right) + \left(\dfrac{3025}{90}\right)}} = \frac{-60}{\sqrt{(11.21 + 33.6)}} = \frac{-60}{6.69} = -8.97$$

Thus, $z = -8.97$ (because we always recommend using a two-tail test, the negative sign is ignored). Our calculated value greatly exceeds the critical value of 1.96, and also 2.58. We have great confidence that there is a statistically significant difference between the sample means, and therefore we reject H_0 at $P < 0.01$, and conclude that urban life probably does affect respiratory function.

Readers may note that the value of 6.69 in the final step of the calculation of z above is the *standard error of the difference*. This is itself a useful number, and its application is described in Section 10.4.

15.11 The *t*-test for comparing the means of two small samples

When samples are small (under about 30 observations in each), a different version of the z-test is called for. The rationale is similar to the z-test, but the formula for t is rather more cumbersome. However, all the terms within it are familiar, being the sample sizes, means and standard deviations:

$$t = \frac{(\bar{x}_1 - \bar{x}_2)}{\sqrt{\left[\frac{(n_1 - 1)s_1^2 + (n_2 - 1)s_2^2}{(n_1 + n_2 - 2)}\right]\left(\frac{n_1 + n_2}{n_1 n_2}\right)}}$$

where the degrees of freedom are $(n_1 + n_2) - 2$.

Example 15.5

The Hamilton Depression Rating (HDR) score is a measure of clinical depression; the higher the score, the greater the degree of depression. Observations of this variable are widely regarded as being normally distributed. In a clinical trial designed to compare real and simulated (sham) electroconvulsive therapy (ECT), observations of HDR were recorded for patients on entry to the study, and again after four weeks of treatment. The primary variable in this study was the percentage change in HDR over the four weeks of treatment for the 'test' and 'control' groups. The observations are recorded in full in Appendix 11. The sample statistics are summarized below:

'Test' group (real ECT) Sample 1: $\bar{x}_1 = 51.92$; $s_1 = 32.41$; $n_1 = 25$
'Control' group (sham ECT)
Sample 2: $\bar{x}_2 = 19.37$; $s_2 = 32.37$; $n_2 = 22$

In other words, in the sample of 25 patients in the 'test' group the score fell by an average of 51.92 HDR units, whilst in the 'control' sample it fell by an average of only 19.37 units. The difference between the samples appears clear, but could it be due to sampling error? The *t*-test is appropriate to the problem. Substituting in the formula for *t*:

$$t = \frac{(51.92 - 19.37)}{\sqrt{\left[\frac{(25-1)1050.4 + (22-1)1047.8}{(25+22-2)}\right]\left(\frac{25+22}{25 \times 22}\right)}}$$

$$= \frac{32.55}{\sqrt{\left[\frac{25209.6 + 22003.8}{45}\right] \times 0.085}} = \frac{32.55}{\sqrt{1049.2 \times 0.085}}$$

$$= \frac{32.55}{9.44} = 3.45$$

(Note that the value of 9.44 in the final step is the *standard error of the difference*.)

Consulting the table of the distribution of *t* (Appendix 2), we find that the calculated value of 3.45 exceeds the tabulated value of about 2.69 at $P = 0.01$ at $(47 - 2) = 45$ df in a two tailed test. (Note: in tables of *t*, the change in the value of *t* after about 30 df is so small that we have to interpolate – that is 'read between' – the value tabulated for 40 and 60 df.)

We reject the null hypothesis, and conclude that there is a statistically significant difference between the sample means. In this trial, at least, there appears to be a marked reduction in HDR score following ECT treatment.

.12   The *t*-test for matched pairs

A specific form of the *t*-test is used when measurements constitute a *matched pair*. As noted in Section 15.2, it is possible to conduct a more sensitive test upon matched data. The rationale for the *t*-test with matched data is similar to that for the *z*- and *t*-tests with unmatched data. As we explained in Section 15.7 in connection with the Wilcoxon test for matched pairs, it is possible to derive a difference, *d*, by subtracting the value of one observation in a matched pair from the other. If the null hypothesis that the samples are drawn from populations with equal means is true, we would expect the mean value of d to be zero. Significant departure from zero can be tested by the t-test for matched pairs according to the formula:

$$t = \frac{\Sigma d}{\sqrt{\dfrac{n\Sigma d^2 - (\Sigma d)^2}{(n-1)}}}$$

where n is the number of matched pairs and $(n-1)$ is the number of degrees of freedom.

Example 15.6

We noted at the end of Section 15.7 that in certain trials it is desirable to eliminate as far as possible variability due to genetic differences. Thus, siblings (brothers and sisters) might be expected to exhibit less genetic variability than individuals selected at random from the population. Pairs of 'identical twins' are ideal for trials in which it is important to minimize possible genetic variation.

In a trial to compare a 'standard' feeding formula with an 'enriched' formula on the growth weight of babies over a standardized time interval, eight pairs of identical twins were selected. One of each pair was selected at random to be given the standard formula. The data are displayed in Table 15.4, which contains a column for the difference (d) between the values of each pair, and another for the squares of the difference (d^2). Values in these columns are summed to give Σd and Σd^2, respectively. Notice that the sign of d is taken into account when summing the column. Which sample is nominated A or B does not affect the eventual value of t. From Σd we calculate $(\Sigma d)^2$ to be $3.05^2 = 9.3025$.

Substituting in the formula for t, with $n = 8$:

$$t = \frac{3.05}{\sqrt{\dfrac{(8 \times 2.4923) - 9.3025}{8 - 1}}} = \frac{3.05}{1.23} = 2.48$$

$t = 2.48$, at 7 degrees of freedom.

Table 15.4 Growth weights (kg) of matched pairs of babies

Sample A: Enriched formula	Sample B: Standard formula	d	d^2
3.70	3.14	0.56	0.3136
2.95	3.1	−0.15	0.0225
3.34	2.61	0.73	0.5329
4.02	3.50	0.52	0.2704
4.32	3.25	1.07	1.1449
3.07	2.63	0.44	0.1936
3.01	3.01	0	0
3.88	4.00	−0.12	0.0144
		$\Sigma d = 3.05$	$\Sigma d^2 = 2.4923$

Consulting the table of the distribution of t (Appendix 2), we find that our calculated value of t at 7 df exceeds the tabulated value of 2.365 at $P = 0.05$, but does not exceed the value of 3.499 at $P = 0.01$ ('two-tailed test' columns). We therefore reject H_0 at the $P = 0.05$ (5%) level, but not at the $P = 0.01$ (1%) level. We can say that the result is statistically 'significant', but not 'highly significant'. It is likely that the investigator would wish to undertake further trials, possibly with a larger number of pairs of twins, before claiming any benefit from the enriched formula.

5.13 Advice on comparing means

1. If there are more than 30 observations in each sample, then use the z-test. If the distribution of data appears to be badly skewed, increase the number of observations, say, to 50. If this is not possible, try transforming the data (Section 9.11) or use the Mann–Whitney U-test.

2. If there are fewer than 30 observations in each sample, use the t-test. This test assumes that sample data are drawn from populations that are normally distributed. If this is clearly not the case, transform the data (Section 9.11) or use the Mann–Whitney U-test.

3. z-tests or t-tests only answer the question: 'Is there a statistically significant difference between the means of two samples?'. They

don't address the more interesting question: '*To what extent* are the means different?' The execution of both z- and t-tests requires the intermediate calculation of the *standard error of the difference*. This statistic may be employed to estimate the difference between the means of the populations from which the samples are drawn (Sections 10.4 and 10.5).

16 ANALYSIS OF VARIANCE – ANOVA

6.1 Why do we need ANOVA?

Chapter 15 discusses ways of comparing the means of two samples. Sometimes, however, we wish to compare the means of more than two samples. Suppose, for example, we have observations of systolic blood pressure measurements from three samples of patients occupying three different wards, A, B and C, in a hospital. It is possible to compare the means of these samples by using z-tests or, if the samples are small, by t-tests. We would need to perform the test three times to compare A-B, A-C and B-C. With the help of a calculator or computer the task is not too daunting. Let us imagine instead that we wish to compare the means of seven samples. In this event, no less than 21 z-tests are required to compare all possible pairs of means. Even if the analyst has sufficient patience to work through this cumbersome treatment, there is an underlying statistical objection to doing so.

We pointed out in Section 11.7 that if the $P = 0.05$ (5 %) level of significance is consistently accepted, a wrong conclusion will be drawn on average once in every 20 tests performed. So, if the means of our imaginary seven samples are compared in 21 z-tests, there is a good chance that at least one false conclusion will be drawn. Of course, the risk of committing a Type 1 error, that is, rejecting H_0 when it should be accepted, is reduced by setting the acceptable significance level to the more stringent one of $P = 0.01$ (1 %). But that increases the risk of making a Type 2 error, namely failing to reject H_0 when it should be rejected. **Analysis of variance (ANOVA)** overcomes these difficulties by allowing comparisons to be made between any number of sample means, all in a single test.

16.2 How ANOVA works

How analysis of *variance* is used to investigate differences between *means* is illustrated in the following example.

Example 16.1

Compare the individual variances of the three samples below with the overall variance when all observations $n = 15$ are aggregated.

Sample 1	Sample 2	Sample 3	Overall
8	9	3	
10	11	5	
12	13	7	
14	15	9	
16	17	11	
$\Sigma x = 60$	$\Sigma x = 65$	$\Sigma x = 35$	$\Sigma x = 160$
$\bar{x} = 12.0$	$\bar{x} = 13.0$	$\bar{x} = 7.00$	$\bar{x} = 10.667$
$s^2 = 10.00$	$s^2 = 10.00$	$s^2 = 10.00$	$s^2 = 16.0$

The means of Samples 1 and 2 are similar; the mean of sample 3 is much lower; the mean of the 15 aggregated observations is intermediate in value, 10.667. The variances of the three are identical (10.00), and therefore the 'average variance' is 10.0. The variance of the aggregated observations, however, is larger (16.0) than the average sample variance. The increase is due to the difference between the *means* of the samples, in particular, the difference between Sample 3 and the other two means. The samples thus give rise to two sources of variability:

1. The variability around each mean *within* a sample (random scatter).

2. The variability *between* the samples due to differences between the means of the populations from which the samples are drawn.

In other words:

$$\text{Variability}_{total} = \text{variability}_{within} + \text{variability}_{between}$$

ANOVA involves the dividing up or **partitioning** the total variability of a number of samples into its components. If the samples are drawn from normally distributed populations with equal means and variances, the *within* variance is the same as the *between* variance. If a statistical test shows that this is not the case, then the samples have been drawn from populations with different means and/or variances. If it assumed that the variances are equal (and this is an underlying assumption in ANOVA), then it is concluded that the discrepancy is due to differences between the *means*. Thus:

H_0 = samples are drawn from normally distributed populations with equal means and variances.

H_1 = population variances are assumed to be equal, and therefore samples are drawn from populations with different means.

There are formal ways for checking that the variances are similar. However, for practical purposes, provided that the variance (s^2) of the largest sample being analysed is not more than about four times that of the smallest sample, you may proceed with the analysis. If you suspect that observations are not approximately normally distributed (e.g. if they are counts of things), then ANOVA may be performed on *transformed* observations (see Section 9.11), or by a suitable non-parametric alternative like the Kruskal–Wallace test (Section 15.5).

When partitioning the total variability of a number of samples, it is simpler to work with *sums of squares* than variances, because adding and subtracting variances is complicated by different numbers of degrees of freedom. However, in the final stages of the analysis, the sums of squares are converted to variances by dividing by the degrees of freedom in order to compare them by means of a test called the *F*-test, that we describe later.

In Section 8.7 we give the formula for estimating the *sums of squares* (SS) as:

$$SS = \Sigma x^2 - \frac{(\Sigma x)^2}{n}$$

The quantity $\frac{(\Sigma x)^2}{n}$ is often referred to as the **correction term (CT)**.

16.3 Procedure for computing ANOVA

Example 16.2

A health care worker browses through patients' records from various wards in a hospital and notices that there appears to be differences between the average systolic blood pressure measurements in different wards. To investigate whether the suspicion has some statistical validity, 10 observations of systolic blood pressure are selected at random from four different wards. The procedure for comparing the means of the four samples is set out as a series of instructions. Please don't be put off by some rather larger numbers during the intermediate stages – it all boils down to a manageable quantity eventually!

1. Cast the data into a table, labelling each Sample 1–4, respectively. Use a scientific calculator to obtain, for each sample, the mean; the standard deviation (square this to obtain the variance); Σx (square this to obtain $(\Sigma x)^2$); and Σx^2. Record this information at the bottom of the column for each sample. At the right-hand side of the table, record the sums of the totals on n, Σx, Σx^2, using the subscript $_T$ to distinguish them from the sample data. These data are presented in Table 16.1.

Table 16.1 Systolic blood pressure measurements from four hospital wards

Ward 1	Ward 2	Ward 3	Ward 4	Total
108	108	109	107	
118	108	103	99	
117	113	109	105	
118	111	105	100	
113	108	107	104	
112	111	108	113	
111	111	110	110	
110	112	108	105	
110	106	113	106	
119	106	114	105	
$n = 10$	$n = 10$	$n = 10$	$n = 10$	$n_T = 40$
$\bar{x} = 113.6$	$\bar{x} = 109.4$	$\bar{x} = 108.6$	$\bar{x} = 105.4$	
$s = 4.03$	$s = 2.50$	$s = 3.31$	$s = 4.14$	
$s^2 = 16.27$	$s^2 = 6.25$	$s^2 = 10.96$	$s^2 = 17.14$	
$\Sigma x = 1136$	$\Sigma x = 1094$	$\Sigma x = 1086$	$\Sigma x = 1054$	$\Sigma x_T = 4370$
$(\Sigma x)^2 = 1290496$	$(\Sigma x)^2 = 1196836$	$(\Sigma x)^2 = 1179396$	$(\Sigma x)^2 = 1110916$	
$\Sigma x^2 = 129196$	$\Sigma x^2 = 119740$	$\Sigma x^2 = 118038$	$\Sigma x^2 = 111246$	$\Sigma x_T^2 = 478220$

2. Calculate a factor called the correction term, CT:

$$CT = \frac{(\Sigma x_T)^2}{n_T} = \frac{(4370)^2}{40} = 477422.5$$

3. Calculate the total sum of squares of the aggregates samples, SS_T:

$$SS_T = \Sigma x_T^2 - CT$$

$$SS_T = 478220 - 477422.5 = 797.5$$

4. Calculate the *between samples* sum of squares, $SS_{between}$

$$SS_{between} = \frac{(\Sigma x_1)^2}{n_1} + \frac{(\Sigma x_2)^2}{n_2} + \frac{(\Sigma x_3)^2}{n_3} + \frac{(\Sigma x_4)^2}{n_4} - CT$$

$$SS_{between} = \frac{1290496}{10} + \frac{1196836}{10} + \frac{1179396}{10} + \frac{1110916}{10} - CT$$

$$SS_{between} = 129049.6 + 119683.6 + 117939.6 + 111091.6 - 477422.5$$

$$SS_{between} = 341.9$$

5. We now need to know the *within samples* sum of squares. Since we already know (Section 16.2) that:

$$SS_T = SS_{between} + SS_{within}$$

we derive SS_{within} simply by subtracting $SS_{between}$ from SS_T:

$$SS_{within} = (SS_T - SS_{between}) = 797.5 - 341.9 = 455.6$$

6. Determine the number of degrees of freedom (df) for each of the calculated values. The rules for determining these are:

df for $SS_T = n_T - 1 = 40 - 1 = 39$
df for $SS_{between} = a - 1$ (where a = number of samples) $= (4 - 1) = 3$
df for $SS_{within} = n_T - a = 40 - 4 = 36$

7. Estimate the variances by dividing each sum of squares by its respective degrees of freedom. (Note: some books call this quantity 'mean
squares' rather than 'variance'):

$$s^2_{between} = \frac{SS_{between}}{df_{between}} = \frac{341.9}{3} = 113.97$$

$$s^2_{between} = \frac{SS_{within}}{df_{within}} = \frac{455.6}{36} = 12.66$$

8. Compute the test statistic F by means of the F-test. The F-test is
extremely simple – divide the *between* variance by the *within* variance
to obtain F:

$$F = \frac{between \; samples \; variance}{within \; samples \; variance} = \frac{113.97}{12.66} = 9.002$$

It should be noted that the denominator (the bottom part) of the
equation is always the *within* samples variance. If it should turn out
that this is larger than the *between* samples variance, then F is less than
1.0. Thus there is no need to compute F, H_0 is automatically accepted.

9. Enter the results in an ANOVA summary table:

Source of variation	SS	df	s^2 ('mean squares')	F
Between	341.9	3	113.97	9.002
Within	455.6	36	12.66	
Total	797.5	39		

Consulting a table of F (Appendix 8), we find that our calculated
value of F at 3 and 36 df exceeds the critical value of 2.88 at the $P = 0.05$
level (interpolating between 30 and 40 df in v_2). We therefore reject the
null hypothesis, and conclude that the variation in the mean blood
pressure between the four hospital wards is statistically significantly
different. Maybe the wards differ according to age or sex categories of
patients. We record the result as:

'The difference in mean systolic blood pressure of the four samples,
where $n = 10$ in each case, is statistically significant ($F_{3, 36} = 9.002$,

$P < 0.05$). (The subscript 3,36 after the F indicates the df of the between and within variances, respectively.)

This is not necessarily the end of the analysis, however. Taking the four means as a group, we know that there is statistically significant variation between them. This may mean that all possible combinations of pairs are different from each other, or that just one is different from the other three. A good indication of which sample means are different from the others is obtained by presenting the individual sample means in histogram form, displaying the 95 percent confidence interval, as shown in Figure 10.1. The samples whose intervals do not overlap are presumed to have been drawn from populations with different means. We suggest you always do this, because it indicates whether the outcome of your ANOVA is 'reasonable'.

A more sensitive test for distinguishing the mean differences that are significantly different is the **Tukey test**. This is simple to apply and is outlined in the next section.

16.4 The Tukey test

The Tukey test only needs to be undertaken when the result of the final F-test in the ANOVA indicates that there is a significant difference between the means of the samples. The simplest version requires that there are an equal number of observations in all samples. The procedure for the test is as follows:

1. Construct a trellis for the comparison of all sample means. This is done below for the blood pressure measurements of Example 16.1. When subtracting one mean from another, ignore any minus signs.

Sample	2	3	4
Sample 1 $\bar{x}_1 = 113.6$	$(\bar{x}_1 - \bar{x}_2)$ 4.2	$(\bar{x}_1 - \bar{x}_3)$ 5.0	$(\bar{x}_1 - \bar{x}_4)$ 8.2
Sample 2 $\bar{x}_2 = 109.4$		$(\bar{x}_2 - \bar{x}_3)$ 0.8	$(\bar{x}_2 - \bar{x}_4)$ 4.0
Sample 3 $\bar{x}_3 = 108.6$			$(\bar{x}_3 - \bar{x}_4)$ 3.2
Sample 4 $\bar{x}_4 = 105.4$			

2. Compute a test statistic T to provide a standard against which the values of the mean differences in the trellis can be compared:

$$T = (q) \times \sqrt{\frac{\text{within variance}}{n}}$$

where $n =$ the number of observations in each sample. The *within variance* is obtained from the ANOVA summary table of Step 9 of the previous section. The value of q is found by consulting the 'Tukey Table' (Appendix 9) for varying numbers of samples, a, and degrees of freedom, v. The respective values in this example are 4 and 36 (36 being the degrees of freedom of the 'within samples' variance). Interpolating Appendix 9 at $a = 4$, $v = 36$, the tabulated value of q is about 3.82 (interpolating between 30 and 40 df).

The T statistic is calculated as:

$$T = 3.82 \times \sqrt{12.66/10} = 4.3$$

There are only two mean differences in the table that exceed this; 1 and 3, and 1 and 4. These two mean differences are therefore significantly different at $P = 0.05$.

The Tukey test above is applied to an example in which there are an equal number of observations (10) in each of the four samples. If all samples do not have the same number of observations, the ANOVA we have described may still be applied; however, there is a modification to the formula used for calculating the critical difference between means using the Tukey test. The formula is:

$$T_{ij} = \frac{q}{\sqrt{2}} \sqrt{\text{within variance}} \times \left(\frac{1}{n_1} + \frac{1}{n_2} \right)$$

T_{ij} is the critical difference between sample i and j which have sample sizes of n_i and n_j, respectively. q is obtained from the Tukey Table (Appendix 9), as before. Note that if the sample sizes are all different, each of these critical differences, T_{ij}, are different.

16.5 Further applications of ANOVA

In Example 16.2, we compared the means of four samples on a measured on single variable (or 'treatment') – namely 'hospital ward'. The four samples were categorized according to a single 'row' of data, and the method is accordingly described as 'one way ANOVA'. Analysis of variance is such a powerful technique that it may be employed to compare sets of samples that are cast into more than 'one row'; in other words, more than one variable (or 'treatment') is considered. Table 16.2 classifies eight samples of systolic blood pressure measurements into two variables ('treatments'): Variable A, gender; and variable B, age group.

The data are clearly classified into two variables (or 'directions' or 'treatments'), and so analysis of variance conducted to compare the sample means in this table is called 'two-way ANOVA'.

There are two particular questions that arise from these data:

1. Is there a significant difference between the mean blood pressures of men and women?

2. Is there a significant difference between the mean blood pressures of the different age groups?

To answer these questions, there are 28 combinations of pairs of means to compare. Whilst it is certainly possible to compare these means by multiple z- or t-tests, the same objections that are raised in Section 16.1 also apply here.

Two-way ANOVA compares all pairs of means simultaneously and involves partitioning the total variability (in the form of 'sums of squares') into the various components that make up the total. Once again, there are two major components of the total sums of squares related in the equation:

$$SS_{total} = SS_{between} + SS_{within}$$

The SS_{within} quantity again represents the variability due to random scatter within each sample. However, in two-way ANOVA, the $SS_{between}$

Table 16.2   Mean (with S.E.) systolic blood pressure measurements (mmHg) of samples of men and women in different age groups

		Variable B, age group			
		20–29	30–39	40–49	50–59
Variable A	Men	Sample 1	Sample 2	Sample 3	Sample 4
		123 ± 14	128 ± 12	133 ± 13	143 ± 14
Gender	Women	Sample 5	Sample 6	Sample 7	Sample 8
		115 ± 13	121 ± 15	130 ± 16	140 ± 15

item is subdivided into the main 'treatment effects' (i.e. the two main variables, age and gender in this example):

1. SS representing variability between samples due to variable A (SS_A).

2. SS representing variability between samples due to variable B (SS_B).

The complete equation is written:

$$SS_{Total} = (SS_A + SS_B) + SS_{within}$$

In two-way ANOVA, all these components are estimated. As before, the analysis concludes by converting the individual values of SS to the corresponding variances by dividing by the respective degrees of freedom. F-tests are conducted to check for significant differences between them.

There is one further possible source of variation that needs to be considered. Scanning the data in Table 16.2, we notice that going along the two rows from left to right, the general trend is the same for both genders – both increase along the row. The trend is roughly 'parallel'. Moreover, moving down from the top row to the bottom, there is a similar parallel trend down the columns – women have generally lower systolic blood pressures than men. When both variables exhibit a 'parallel trend' like this, we say that the variables are **additive**, and that there is no **interaction** between the variables. In this example, we can predict negligible contribution to the total variability due to interaction.

On the other hand, if the two variables do not exhibit a 'parallel trend', for instance if one gender's values increased with age whilst the other's decreased, there could be a significant contribution to the 'between samples' sum of squares by virtue of an **interaction sum of squares** SS_I. The equation for total variability is modified accordingly:

$$SS_T = (SS_A + SS_B + SS_I) + SS_{within}$$

A full two-way analysis of variance estimates all these quantities. However, such a full analysis is beyond the scope of this introductory book. Readers who consider that they are at a stage to tackle this more advanced treatment are invited to consult the companion text by Fowler, Cohen and Jarvis (1998) for a fully worked example. Here, the variables concern the weights of samples of birds, but having got this far we anticipate that you will have no difficulty in adapting the example to deal with variables of personal interest!

16.6 Advice on using ANOVA

ANOVA is an efficient and powerful technique for investigating relationships between samples. There are, however, a number of restrictions governing the use of the method:

1. ANOVA assumes that all observations are obtained randomly and are normally distributed. Certain data may be *transformed* to normal, as described in Section 9.11.

2. ANOVA assumes that the variances of the sample are similar. If the largest variance among the samples is more than about four times the smallest, it may be necessary to first transform the data.

3. In one-way ANOVA, sample sizes do not have to be equal. When the result on ANOVA indicates significant differences between one or more means, Tukey's test can be used to pin-pint which means are different from which.

APPENDICES

Appendix 1: Table of random numbers

23157	54859	01837	25993	76249	70886	95230	36744
05545	55043	10537	43508	90611	83744	10962	21343
14871	60350	32404	36223	50051	00322	11543	80834
38976	74951	94051	75853	78805	90194	32428	71695
97312	61718	99755	30870	94251	25841	54882	10513
11742	69381	44339	30872	32797	33118	22647	06850
43361	28859	11016	45623	93009	00499	43640	74036
93806	20478	38268	04491	55751	18932	58475	52571
49540	13181	08429	84187	69538	29661	77738	09527
36768	72633	37948	21569	41959	68670	45274	83880
07092	52392	24627	12067	06558	45344	67338	45320
43310	01081	44863	80307	52555	16148	89742	94647
61570	06360	06173	63775	63148	95123	35017	46993
31352	83799	10779	18941	31579	76448	62584	86919
57048	86526	27795	93692	90529	56546	35065	32254
09243	44200	68721	07137	30729	75756	09298	27650
97957	35018	40894	88329	52230	82521	22532	61587
93732	59570	43781	98885	56671	66826	95996	44569
72621	11225	00922	68264	35666	59434	71687	58167
61020	74418	45371	20794	95917	37866	99536	19378
97839	85474	33055	91718	45473	54144	22034	23000
89160	97192	22232	90637	35055	45489	88438	16361
25966	88220	62871	79265	02823	52862	84919	54883
81443	31719	05049	54806	74690	07567	65017	16543
11322	54931	42362	34386	08624	97687	46245	23245

Appendix 2: *t*-distribution

d.f.	Level of significance for one-tailed test			
	0.05	0.025	0.01	0.005
	Level of significance for two-tailed test			
	0.10	0.05	0.02	0.01
1	6.314	12.706	31.821	63.657
2	2.920	4.303	6.965	9.925
3	2.353	3.182	4.541	5.841
4	2.132	2.776	3.747	4.604
5	2.015	2.571	3.365	4.032
6	1.943	2.447	3.143	3.707
7	1.895	2.365	2.998	3.499
8	1.860	2.306	2.896	3.355
9	1.833	2.262	2.821	3.250
10	1.812	2.228	2.764	3.169
11	1.796	2.201	2.718	3.106
12	1.782	2.179	2.681	3.055
13	1.771	2.160	2.650	3.012
14	1.761	2.145	2.624	2.977
15	1.753	2.131	2.602	2.947
16	1.746	2.120	2.583	2.921
17	1.740	2.110	2.567	2.898
18	1.734	2.101	2.552	2.878
19	1.729	2.093	2.539	2.861
20	1.725	2.086	2.528	2.845
21	1.721	2.080	2.518	2.831
22	1.717	2.074	2.508	2.819
23	1.714	2.069	2.500	2.807
24	1.711	2.064	2.492	2.797
25	1.708	2.060	2.485	2.787
26	1.706	2.056	2.479	2.779
27	1.703	2.052	2.473	2.771
28	1.701	2.048	2.467	2.763
29	1.699	2.045	2.462	2.756
30	1.697	2.042	2.457	2.750
40	1.684	2.021	2.423	2.704
60	1.671	2.000	2.390	2.660
120	1.658	1.980	2.358	2.617
∞	1.645	1.960	2.326	2.576

Appendix 3: χ^2-distribution

Degrees of freedom	Level of significance	
	0.05	0.01
1	3.84	6.63
2	5.99	9.21
3	7.81	11.34
4	9.49	13.28
5	11.07	15.09
6	12.59	16.81
7	14.07	18.48
8	15.51	20.09
9	16.92	21.67
10	18.31	23.21
11	19.68	24.72
12	21.03	26.22
13	22.36	27.69
14	23.68	29.14
15	25.00	30.58
16	26.30	32.00
17	27.59	33.41
18	28.87	34.81
19	30.14	36.19
20	31.41	37.57
21	32.67	38.93
22	33.92	40.29
23	35.17	41.64
24	36.42	42.98
25	37.65	44.31
26	38.89	45.64
27	40.11	46.96
28	41.34	48.28
29	42.56	49.59
30	43.77	50.89
40	55.76	63.69
50	67.50	76.15
60	79.08	88.38
70	90.53	100.43
80	101.88	112.33
90	113.15	124.12
100	124.34	135.81

Appendix 4: Critical values of Spearman's Rank Correlation Coefficient

	Level of significance for one-tailed test			
	0.05	0.025	0.01	0.005
	Level of significance for two-tailed test			
n	0.10	0.05	0.02	0.01
5	0.900	—	—	
6	0.829	0.886	0.943	—
7	0.714	0.786	0.893	—
8	0.643	0.738	0.833	0.881
9	0.600	0.683	0.783	0.833
10	0.564	0.648	0.745	0.794
11	0.523	0.623	0.736	0.818
12	0.497	0.591	0.703	0.780
13	0.475	0.566	0.673	0.745
14	0.457	0.545	0.646	0.716
15	0.441	0.525	0.623	0.689
16	0.425	0.507	0.601	0.666
17	0.412	0.490	0.582	0.645
18	0.399	0.476	0.564	0.625
19	0.388	0.462	0.549	0.608
20	0.377	0.450	0.534	0.591
21	0.368	0.438	0.521	0.576
22	0.359	0.428	0.508	0.562
23	0.351	0.418	0.496	0.549
24	0.343	0.409	0.485	0.537
25	0.336	0.400	0.475	0.526
26	0.329	0.392	0.465	0.515
27	0.323	0.385	0.456	0.505
28	0.317	0.377	0.448	0.496
29	0.311	0.370	0.440	0.487
30	0.305	0.364	0.432	0.478

Appendix 5: Product moment correlation values at the 0.05 and 0.01 levels of significance

d.f.	0.05	0.01
1	0.997	0.9999
2	0.950	0.990
3	0.878	0.959
4	0.811	0.917
5	0.754	0.874
6	0.707	0.834
7	0.666	0.798
8	0.632	0.765
9	0.602	0.735
10	0.576	0.708
11	0.553	0.684
12	0.532	0.661
13	0.514	0.641
14	0.497	0.623
15	0.482	0.606
16	0.468	0.590
17	0.456	0.575
18	0.444	0.561
19	0.433	0.549
20	0.423	0.537
21	0.413	0.526
22	0.404	0.515
23	0.396	0.505
24	0.388	0.496
25	0.381	0.487
26	0.374	0.479
27	0.367	0.471
28	0.361	0.463
29	0.355	0.456
30	0.349	0.449
32	0.339	0.436
34	0.329	0.424

d.f.	0.05	0.01
35	0.325	0.418
36	0.320	0.413
38	0.312	0.403
40	0.304	0.393
42	0.297	0.384
44	0.291	0.376
45	0.288	0.372
46	0.284	0.368
48	0.279	0.361
50	0.273	0.354
55	0.261	0.338
60	0.250	0.325
65	0.241	0.313
70	0.232	0.302
75	0.224	0.292
80	0.217	0.283
85	0.211	0.275
90	0.205	0.267
95	0.200	0.260
100	0.195	0.254
125	0.174	0.228
150	0.159	0.208
175	0.148	0.193
200	0.138	0.181
300	0.113	0.148
400	0.098	0.128
500	0.088	0.115
1000	0.062	0.081

Appendix 6: Mann–Whitney U-test values (two-tailed test) P = 0.05

n_1\\n_2	2	3	4	5	6	7	8	9	10	11	12	13	14	15	16	17	18	19	20
2							0	0	0	0	1	1	1	1	1	2	2	2	2
3				0	1	1	2	2	3	3	4	4	5	5	6	6	7	7	8
4			0	1	2	3	4	4	5	6	7	8	9	10	11	11	12	13	13
5		0	1	2	3	5	6	7	8	9	11	12	13	14	15	17	18	19	20
6		1	2	3	5	6	8	10	11	13	14	16	17	19	21	22	24	25	27
7		1	3	5	6	8	10	12	14	16	18	20	22	24	26	28	30	32	34
8	0	2	4	6	8	10	13	15	17	19	22	24	26	29	31	34	36	38	41
9	0	2	4	7	10	12	15	17	20	23	26	28	31	34	37	39	42	45	48
10	0	3	5	8	11	14	17	20	23	26	29	33	36	39	42	45	48	52	55
11	0	3	6	9	13	16	19	23	26	30	33	37	40	44	47	51	55	58	62
12	1	4	7	11	14	18	22	26	29	33	37	41	45	49	53	57	61	65	69
13	1	4	8	12	16	20	24	28	33	37	41	45	50	54	59	63	67	72	76
14	1	5	9	13	17	22	26	31	36	40	45	50	55	59	64	67	74	78	83
15	1	5	10	14	19	24	29	34	39	44	49	54	59	64	70	75	80	85	90
16	1	6	11	15	21	26	31	37	42	47	53	59	64	70	75	81	86	92	98
17	2	6	11	17	22	28	34	39	45	51	57	63	67	75	81	87	93	99	105
18	2	7	12	18	24	30	36	42	48	55	61	67	74	80	86	93	99	106	112
19	2	7	13	19	25	32	38	45	52	58	65	72	78	85	92	99	106	113	119
20	2	8	13	20	27	34	41	48	55	62	69	76	83	90	98	105	112	119	127

n_1 and n_2 are the number of observations in each sample

Appendix 7: Critical values of *T* in the Wilcoxon test for two matched samples

Sample size	Levels of significance			
	One-tailed test			
	0.05	0.025	0.01	0.001
	Two-tailed test			
	0.1	0.05	0.02	0.002
N = 5	T ≤ 0			
6	2	0		
7	3	2	0	
8	5	3	1	
9	8	5	3	
10	10	8	5	0
11	13	10	7	1
12	17	13	9	2
13	21	17	12	4
14	25	21	15	6
15	30	25	19	8
16	35	29	23	11
17	41	34	27	14
18	47	40	32	18
19	53	46	37	21
20	60	52	43	26
21	67	58	49	30
22	75	65	55	35
23	83	73	62	40
24	91	81	69	45
25	100	89	76	51
26	110	98	84	58
27	119	107	92	64
28	130	116	101	71
30	151	137	120	86
31	163	147	130	94
32	175	159	140	103
33	187	170	151	112

Appendix 8: *F*-distribution

Use these tables for testing significance in analysis of variance. (a) 0.05
level; (b) 0.01 level

v_1 = df for the greater variance
v_2 = df for the lesser variance

(*a*)

v_2 \ v_1	1	2	3	4	5	6	7	8	9
1	161.45	199.50	215.71	224.58	230.16	233.99	236.77	238.88	240.54
2	18.513	19.000	19.164	19.247	19.296	19.330	19.353	19.371	19.385
3	10.128	9.5521	9.2766	9.1172	9.0135	8.9406	8.8867	8.8452	8.8323
4	7.7086	6.9443	6.5914	6.3882	6.2561	6.1631	6.0942	6.0410	5.9938
5	6.6079	5.7861	5.4095	5.1922	5.0503	4.9503	4.8759	4.8183	4.7725
6	5.9874	5.1433	4.7571	4.5337	4.3874	4.2839	4.2067	4.1468	4.0990
7	5.5914	4.7374	4.3468	4.1203	3.9715	3.8660	3.7870	3.7257	3.6767
8	5.3177	4.4590	4.0662	3.8379	3.6875	3.5806	3.5005	3.4381	3.3881
9	5.1174	4.2565	3.8625	3.6331	3.4817	3.3738	3.2927	3.2296	3.1789
10	4.9646	4.1028	3.7083	3.4780	3.3258	3.2172	3.1355	3.0717	3.0204
11	4.8443	3.9823	3.5874	3.3567	3.2039	3.0946	3.0123	2.9480	2.8962
12	4.7472	3.8853	3.4903	3.2592	3.1059	2.9961	2.9134	2.8486	2.7964
13	4.6672	3.8056	3.4105	3.1791	3.0254	2.9153	2.8321	2.7669	2.7444
14	4.6001	3.7389	3.3439	3.1122	2.9582	2.8477	2.7642	2.6987	2.6458
15	4.5431	3.6823	3.2874	3.0556	2.9013	2.7905	2.7066	2.6408	2.5876
16	4.4940	3.6337	3.2389	3.0069	2.8524	2.7413	2.6572	2.5911	2.5377
17	4.4513	3.5915	3.1968	2.9647	2.8100	2.6987	2.6143	2.5480	2.4443
18	4.4139	3.5546	3.1599	2.9277	2.7729	2.6613	2.5767	2.5102	2.4563
19	4.3807	3.5219	3.1274	2.8951	2.7401	2.6283	2.5435	2.4768	2.4227
20	4.3512	3.4928	3.0984	2.8661	2.7109	2.5990	2.5140	2.4471	2.3928
21	4.3248	3.4668	3.0725	2.8401	2.6848	2.5727	2.4876	2.4205	2.3660
22	4.3009	3.4434	3.0491	2.8167	2.6613	2.5491	2.4638	2.3965	2.3219
23	4.2793	3.4221	3.0280	2.7955	2.6400	2.5277	2.4422	2.3748	2.3201
24	4.2597	3.4028	3.0088	2.7763	2.6207	2.5082	2.4226	2.3551	2.3002
25	4.2417	3.3852	2.9912	2.7587	2.6030	2.4904	2.4047	2.3371	2.2821
26	4.2252	3.3690	2.9752	2.7426	2.5868	2.4741	2.3883	2.3205	2.2655
27	4.2100	3.3541	2.9604	2.7278	2.5719	2.4591	2.3732	2.3053	2.2501
28	4.1960	3.3404	2.9467	2.7141	2.5581	2.4453	2.3593	2.2913	2.2360
29	4.1830	3.3277	2.9340	2.7014	2.5454	2.4324	2.3463	2.2783	2.2329
30	4.1709	3.3158	2.9223	2.6896	2.5336	2.4205	2.3343	2.2662	2.2507
40	4.0847	3.2317	2.8387	2.6060	2.4495	2.3359	2.2490	2.1802	2.1240
60	4.0012	3.1504	2.7581	2.5252	2.3683	2.2541	2.1665	2.0970	2.0401
120	3.9201	3.0718	2.6802	2.4472	2.2899	2.1750	2.0868	2.0164	1.9688
∞	3.8415	2.9957	2.6049	2.3719	2.2141	2.0986	2.0096	1.9384	1.8799

Appendix 8 (*cont.*)

10	12	15	20	24	30	40	60	120	∞
241.88	243.91	245.95	248.01	249.05	250.10	251.14	252.20	253.25	254.31
19.396	19.413	19.429	19.446	19.454	19.462	19.471	19.479	19.487	19.496
8.7855	8.7446	8.7029	8.6602	8.6385	8.6166	8.5944	8.5720	8.5594	8.5264
5.9644	5.9117	5.8578	5.8025	5.7744	5.7459	5.7170	5.6877	5.6381	5.6281
4.7351	4.6777	4.6188	4.5581	4.5272	4.4957	4.4638	4.4314	4.3085	4.3650
4.0600	3.9999	3.9381	3.8742	3.8415	3.8082	3.7743	3.7398	3.7047	3.6689
3.6365	3.5747	3.5107	3.4445	3.4105	3.3758	3.3404	3.3043	3.2674	3.2298
3.3472	3.2839	3.2184	3.1503	3.1152	3.0794	3.0428	3.0053	2.9669	2.9276
3.1373	3.0729	3.0061	2.9365	2.9005	2.8637	2.8259	2.7872	2.7475	2.7067
2.9782	2.9130	2.8450	2.7740	2.7372	2.6996	2.6609	2.6211	2.5801	2.5379
2.8536	2.7876	2.7186	2.6464	2.6090	2.5705	2.5309	2.4901	2.4480	2.4045
2.7534	2.6866	2.6169	2.5436	2.5055	2.4663	2.4259	2.3842	2.3410	2.2962
2.6710	2.6037	2.5331	2.4589	2.4202	2.3803	2.3392	2.2966	2.2524	2.2064
2.6022	2.5342	2.4630	2.3879	2.3487	2.3082	2.2664	2.2229	2.1778	2.1307
2.5437	2.4753	2.4034	2.3275	2.2878	2.2468	2.2043	2.1601	2.1141	2.0658
2.4935	2.4247	2.3522	2.2756	2.2354	2.1938	2.1507	2.1058	2.0589	2.0096
2.4499	2.3807	2.3077	2.2304	2.1898	2.1477	2.1040	2.0584	2.0107	1.9604
2.4117	2.3421	2.2686	2.1906	2.1497	2.1071	2.0629	2.0166	1.9681	1.9168
2.3779	2.3080	2.2341	2.1555	2.1141	2.0712	2.0264	1.9795	1.9302	1.8780
2.3479	2.2776	2.2033	2.1242	2.0825	2.0391	1.9938	1.9464	1.8963	1.8432
2.3210	2.2504	2.1757	2.0960	2.0540	2.0102	1.9645	1.9165	1.8657	1.8117
2.2967	2.2258	2.1508	2.0707	2.0283	1.9842	1.9380	1.8894	1.8380	1.7831
2.2747	2.2036	2.1282	2.0476	2.0050	1.9605	1.9139	1.8648	1.8128	1.7570
2.2547	2.1834	2.1077	2.0267	1.9838	1.9390	1.8920	1.8424	1.7896	1.7330
2.2365	2.1649	2.0889	2.0075	1.9643	1.9192	1.8718	1.8217	1.7684	1.7110
2.2197	2.1479	2.0716	1.9898	1.9464	1.9010	1.8533	1.8027	1.7488	1.6906
2.2043	2.1323	2.0558	1.9736	1.9299	1.8842	1.8361	1.7851	1.7306	1.6717
2.1900	2.1179	2.0411	1.9586	1.9147	1.8687	1.8203	1.7689	1.7138	1.6541
2.1768	2.1045	2.0275	1.9446	1.9005	1.8543	1.8055	1.7537	1.6981	1.6376
2.1646	2.0921	2.0148	1.9317	1.8874	1.8409	1.7918	1.7396	1.6835	1.6223
2.0772	2.0035	1.9245	1.8389	1.7929	1.7444	1.6928	1.6373	1.5766	1.5089
1.9926	1.9174	1.8364	1.7480	1.7001	1.6491	1.5943	1.5343	1.4673	1.3893
1.9105	1.8337	1.7505	1.6587	1.6084	1.5543	1.4952	1.4290	1.3519	1.2539
1.8307	1.7522	1.6664	1.5705	1.5173	1.4591	1.3940	1.3180	1.0214	1.0000

Appendix 8 (*cont.*)

(*b*)

v_1 / v_2	1	2	3	4	5	6	7	8	9
1	4052.2	4999.5	5403.4	5624.6	5763.6	5859.0	5928.4	5981.1	6022.5
2	98.503	99.000	99.166	99.249	99.299	99.333	99.356	99.374	99.388
3	34.116	30.817	29.457	28.710	28.237	27.911	27.672	27.489	27.345
4	21.198	18.000	16.694	15.977	15.522	15.207	14.976	14.799	14.659
5	16.258	13.274	12.060	11.392	10.967	10.672	10.456	10.289	10.158
6	13.745	10.925	9.7795	9.1483	8.7459	8.4661	8.2600	8.1017	7.9761
7	12.246	9.5466	8.4513	7.8466	7.4604	7.1914	6.9928	6.8400	6.7188
8	11.259	8.6491	7.5910	7.0061	6.6318	6.3707	6.1776	6.0289	5.9106
9	10.561	8.0215	6.9919	6.4221	6.0569	5.8018	5.6129	5.4671	5.3511
10	10.044	7.5594	6.5523	5.9943	5.6363	5.3858	5.2001	5.0567	4.9424
11	9.6460	7.2057	6.2167	5.6683	5.3160	5.0692	4.8861	4.7445	4.6315
12	9.3302	6.9266	5.9525	5.4120	5.0643	4.8206	4.6395	4.4994	4.3875
13	9.0738	6.7010	5.7394	5.2053	4.8616	4.6204	4.4410	4.3021	4.1911
14	8.8616	6.5149	5.5639	5.0354	4.6950	4.4558	4.2779	4.1399	4.0297
15	8.6831	6.3589	5.4170	4.8932	4.5556	4.3183	4.1415	4.0045	3.8948
16	8.5310	6.2262	5.2922	4.7726	4.4374	4.2016	4.0259	3.8896	3.7804
17	8.3997	6.1121	5.1850	4.6690	4.3359	4.1015	3.9267	3.7910	3.6822
18	8.2854	6.0129	5.0919	4.5790	4.2479	4.0146	3.8406	3.7054	3.5971
19	8.1849	5.9259	5.0103	4.5003	4.1708	3.9386	3.7653	3.6305	3.5225
20	8.0960	5.8489	4.9382	4.4307	4.1027	3.8714	3.6987	3.5644	3.4567
21	8.0166	5.7804	4.8740	4.3688	4.0421	3.8117	3.6396	3.5056	3.3981
22	7.9454	5.7190	4.8166	4.3134	3.9880	3.7583	3.5867	3.4530	3.3458
23	7.8811	5.6637	4.7649	4.2636	3.9392	3.7102	3.5390	3.4057	3.2986
24	7.8229	5.6136	4.7181	4.2184	3.8951	3.6667	3.4959	3.3629	3.2560
25	7.7698	5.5680	4.6755	4.1774	3.8550	3.6272	3.4568	3.3239	3.2172
26	7.7213	5.5263	4.6366	4.1400	3.8183	3.5911	3.4210	3.2884	3.1818
27	7.6767	5.4881	4.6009	4.1056	3.7848	3.5580	3.3882	3.2558	3.1494
28	7.6356	5.4529	4.5681	4.0740	3.7539	3.5276	3.3581	3.2259	3.1195
29	7.5977	5.4204	4.5378	4.0449	3.7254	3.4995	3.3303	3.1982	3.0920
30	7.5625	5.3903	4.5097	4.0179	3.6990	3.4735	3.3045	3.1726	3.0665
40	7.3141	5.1785	4.3126	3.8283	3.5138	3.2910	3.1238	2.9930	2.8876
60	7.0771	4.9774	4.1259	3.6490	3.3389	3.1187	2.9530	2.8233	2.7185
120	6.8509	4.7865	3.9491	3.4795	3.1735	2.9559	2.7918	2.6629	2.5586
∞	6.6349	4.6052	3.7816	3.3192	3.0173	2.8020	2.6393	2.5113	2.4073

Appendix 8 (*cont.*)

10	12	15	20	24	30	40	60	120	∞
6055.8	6106.3	6157.3	6208.7	6234.6	6260.6	6286.8	6313.0	6339.4	6365.9
99.399	99.416	99.433	99.449	99.458	99.466	99.474	99.482	99.491	99.499
27.229	27.052	26.872	26.690	26.598	26.505	26.411	26.316	26.221	26.125
14.546	14.374	14.198	14.020	13.929	13.838	13.745	13.652	13.558	13.463
10.051	9.8883	9.7222	9.5526	9.4665	9.3793	9.2912	9.2020	9.1118	9.0204
7.8741	7.7183	7.5590	7.3958	7.3127	7.2285	7.1432	7.0567	6.9690	6.8800
6.6201	6.4691	6.3143	6.1554	6.0743	5.9920	5.9084	5.8236	5.7373	5.6495
5.8143	5.6667	5.5151	5.3591	5.2793	5.1981	5.1156	5.0316	4.9461	4.8588
5.2565	5.1114	4.9621	4.8080	4.7290	4.6486	4.5666	4.4831	4.3978	4.3105
4.8491	4.7059	4.5581	4.4054	4.3269	4.2469	4.1653	4.0819	3.9965	3.9090
4.5393	4.3974	4.2509	4.0990	4.0209	3.9411	3.8596	3.7761	3.6904	3.6024
4.2961	4.1553	4.0096	3.8584	3.7805	3.7008	3.6192	3.5355	3.4494	3.3608
4.1003	3.9603	3.8154	3.6646	3.5868	3.5070	3.4253	3.3413	3.2548	3.1654
3.9394	3.8001	3.6557	3.5052	3.4274	3.3476	3.2656	3.1813	3.0942	3.0040
3.8049	3.6662	3.5222	3.3719	3.2940	3.2141	3.1319	3.0471	2.9595	2.8684
3.6909	3.5527	3.4089	3.2587	3.1808	3.1008	3.0182	2.9330	2.8447	2.7528
3.5931	3.4552	3.3117	3.1615	3.0835	3.0032	2.9205	2.8348	2.7459	2.6530
3.5082	3.3706	3.2273	3.0771	2.9990	2.9185	2.8354	2.7493	2.6597	2.5660
3.4338	3.2965	3.1533	3.0031	2.9249	2.8442	2.7608	2.6742	2.5839	2.4893
3.3682	3.2311	3.0880	2.9377	2.8594	2.7785	2.6947	2.6077	2.5168	2.4212
3.3098	3.1730	3.0300	2.8796	2.8010	2.7200	2.6359	2.5484	2.4568	2.3603
3.2576	3.1209	2.9779	2.8274	2.7488	2.6675	2.5831	2.4951	2.4029	2.3055
3.2106	3.0740	2.9311	2.7805	2.7017	2.6202	2.5355	2.4471	2.3542	2.2558
3.1681	3.0316	2.8887	2.7380	2.6591	2.5773	2.4923	2.4035	2.3100	2.2107
3.1294	2.9931	2.8502	2.6993	2.6203	2.5383	2.4530	2.3637	2.2696	2.1694
3.0941	2.9578	2.8150	2.6640	2.5848	2.5026	2.4170	2.3273	2.2325	2.1315
3.0618	2.9256	2.7827	2.6316	2.5522	2.4699	2.3840	2.2938	2.1985	2.0965
3.0320	2.8959	2.7530	2.6017	2.5223	2.4397	2.3535	2.2629	2.1670	2.0642
3.0045	2.8685	2.7256	2.5742	2.4946	2.4118	2.3253	2.2344	2.1379	2.0342
2.9791	2.8431	2.7002	2.5487	2.4689	2.3860	2.2992	2.2079	2.1108	2.0062
2.8005	2.6648	2.5216	2.3689	2.2880	2.2034	2.1142	2.0194	1.9172	1.8047
2.6318	2.4961	2.3523	2.1978	2.1154	2.0285	1.9360	1.8363	1.7263	1.6006
2.4721	2.3363	2.1915	2.0346	1.9500	1.8600	1.7628	1.6557	1.5330	1.3805
2.3209	2.1847	2.0385	1.8783	1.7908	1.6964	1.5923	1.4730	1.3246	1.0000

Appendix 9: Tukey test

$(p = 0.05)$

a = the total number of means being compared

u = degrees of freedom of denominator of F test

v \ a	2	3	4	5	6	7	8	9	10
1	17.97	26.98	32.82	37.08	40.41	43.12	45.40	47.36	49.07
2	6.08	8.33	9.80	10.88	11.74	12.44	13.03	13.54	13.99
3	4.50	5.91	6.82	7.50	8.04	8.48	8.85	9.18	9.46
4	3.93	5.04	5.76	6.29	6.71	7.05	7.35	7.60	7.83
5	3.64	4.60	5.22	5.67	6.03	6.33	6.58	6.80	6.99
6	3.46	4.34	4.90	5.30	5.63	5.90	6.12	6.32	6.49
7	3.34	4.16	4.68	5.06	5.36	5.61	5.82	6.00	6.16
8	3.26	4.04	4.53	4.89	5.17	5.40	5.60	5.77	5.92
9	3.20	3.95	4.41	4.76	5.02	5.24	5.43	5.59	5.74
10	3.15	3.88	4.33	4.65	4.91	5.12	5.30	5.46	5.60
11	3.11	3.82	4.26	4.57	4.82	5.03	5.20	5.35	5.49
12	3.08	3.77	4.20	4.51	4.75	4.95	5.12	5.27	5.39
13	3.06	3.73	4.15	4.45	4.69	4.88	5.05	5.19	5.32
14	3.03	3.70	4.11	4.41	4.64	4.83	4.99	5.13	5.25
15	3.01	3.67	4.08	4.37	4.59	4.78	4.94	5.08	5.20
16	3.00	3.65	4.05	4.33	4.56	4.74	4.90	5.03	5.15
17	2.98	3.63	4.02	4.30	4.52	4.70	4.86	4.99	5.11
18	2.97	3.61	4.00	4.28	4.49	4.67	4.82	4.96	5.07
19	2.96	3.59	3.98	4.25	4.47	4.65	4.79	4.92	5.04
20	2.95	3.58	3.96	4.23	4.45	4.62	4.77	4.90	5.01
24	2.92	3.53	3.90	4.17	4.37	4.54	4.68	4.81	4.92
30	2.89	3.49	3.85	4.10	4.30	4.46	4.60	4.72	4.82
40	2.86	3.44	3.79	4.04	4.23	4.39	4.52	4.63	4.73
60	2.83	3.40	3.74	3.98	4.16	4.31	4.44	4.55	4.65
120	2.80	3.36	3.68	3.92	4.10	4.24	4.36	4.47	4.56
∞	2.77	3.31	3.63	3.86	4.03	4.17	4.29	4.39	4.47

Appendix 9 (*cont.*)

a / v	11	12	13	14	15	16	17	18	19	20
1	50.59	51.96	53.20	54.33	55.36	56.32	57.22	58.04	58.83	59.56
2	14.39	14.75	15.08	15.38	15.65	15.91	16.14	16.37	16.57	16.77
3	9.72	9.95	10.15	10.35	10.52	10.69	10.84	10.98	11.11	11.24
4	8.03	8.21	8.37	8.52	8.66	8.79	8.91	9.03	9.13	9.23
5	7.17	7.32	7.47	7.60	7.72	7.83	7.93	8.03	8.12	8.21
6	6.65	6.79	6.92	7.03	7.14	7.24	7.34	7.43	7.51	7.59
7	6.30	6.43	6.55	6.66	6.76	6.85	6.94	7.02	7.10	7.17
8	6.05	6.18	6.29	6.39	6.48	6.57	6.65	6.73	6.80	6.87
9	5.87	5.98	6.09	6.19	6.28	6.36	6.44	6.51	6.58	6.64
10	5.72	5.83	5.93	6.03	6.11	6.19	6.27	6.34	6.40	6.47
11	5.61	5.71	5.81	5.90	5.98	6.06	6.13	6.20	6.27	6.33
12	5.51	5.61	5.71	5.80	5.88	5.95	6.02	6.09	6.15	6.21
13	5.43	5.53	5.63	5.71	5.79	5.86	5.93	5.99	6.05	6.11
14	5.36	5.46	5.55	5.64	5.71	5.79	5.85	5.91	5.97	6.03
15	5.31	5.40	5.49	5.57	5.65	5.72	5.78	5.85	5.90	5.96
16	5.26	5.35	5.44	5.52	5.59	5.66	5.73	5.79	5.84	5.90
17	5.21	5.31	5.39	5.47	5.54	5.61	5.67	5.73	5.79	5.84
18	5.17	5.27	5.35	5.43	5.50	5.57	5.63	5.69	5.74	5.79
19	5.14	5.23	5.31	5.39	5.46	5.53	5.59	5.65	5.70	5.75
20	5.11	5.20	5.28	5.36	5.43	5.49	5.55	5.61	5.66	5.71
24	5.01	5.10	5.18	5.25	5.32	5.38	5.44	5.49	5.55	5.59
30	4.92	5.00	5.08	5.15	5.21	5.27	5.33	5.38	5.43	5.47
40	4.82	4.90	4.98	5.04	5.11	5.16	5.22	5.27	5.31	5.36
60	4.73	4.81	4.88	4.94	5.00	5.06	5.11	5.15	5.20	5.24
120	4.64	4.71	4.78	4.84	4.90	4.95	5.00	5.04	5.09	5.13
∞	4.55	4.62	4.68	4.74	4.80	4.85	4.89	4.93	4.97	5.01

Appendix 10: Symbols

The following symbols are the ones we have adopted for use. Whilst most of them are in general use, variations are to be found in statistical literature.

$<$	less than: $3<4$.
$>$	more than: $2>1$.
x	the numerical value of an observation: e.g. wing length x mm; also the value of a frequency class.
f	frequency; the number of observations in a frequency class x.
y	the numerical value of an observation of a second variable from the same sampling unit from which x is taken.
x^2	x squared.
$\sqrt{x}$	square root of x.
N	number of sampling units in a population.
n	number of sampling units in a sample.
n_i	the ith observation in a series of n observations.
Σ	capital sigma: the sum of.
Σx	the sum of all values of x in a series of n observations.
$(\Sigma x)^2$	the square of the sum of x in a series of n observations.
Σx^2	the sum of the squares of x in a series of n observations.
μ	mu: population mean.
$\bar{x}$	x bar: sample mean of n values of x.
$\bar{x}'$	x bar primed: derived mean obtained from transformed values of x.
$\bar{y}$	y bar: sample mean of n values of y,
σ	sigma: population standard deviation.
s	estimate of σ from sample data.
σ^2	sigma squared: population variance.
s^2	estimate of σ^2 from sample data.
P	probability.
z	standard deviation unit of the normal curve; test statistic in the z-test.
ν	nu: degrees of freedom.
H_0	Null hypothesis.
H_1	alternative to a null hypothesis.
χ^2	chi square: test statistic of the chi-square test.
r	product moment correlation coefficient of a sample.
ρ	rho: product moment correlation coefficient of a population.
r^2	coefficient of determination.
r_s	Spearman rank correlation coefficient.
a	intercept of a regression line on the y-axis; number of samples being compared.
b	gradient of a regression line (also known as the regression coefficient).
t	test statistic of the t-test.
U	test statistic of the Mann–Whitney U-test.
T	Test statistic of the Wilcoxon's test for matched pairs.
F	test statistic of the F test.
q	test statistic of the Tukey test.

Appendix 11: Leicestershire ECT study data: Subgroup with depressive illness

		Week															
		0				2				4				12			
Case	Group	Mass	hdr	va1	va2	Mass	hdr	va1	va2	Mass	hdr	va1	va2	Mass	hdr	va1	va2
2	1	5	52	90	72	7	36	68	61	—	—	97	87	11	54	89	80
4	1	5	36	64	32	1	4	55	11	—	—	—	—	5	36	58	25
7	1	2	38	60	22	0	0	42	1	0	0	47	0	1	2	55	14
8	1	—	—	97	97	—	—	96	95	1	2	51	3	1	8	50	9
10	0	8	40	93	80	—	—	95	89	5	34	97	89	—	—	90	85
11	0	4	22	95	80	5	26	95	91	3	50	96	93	2	6	62	16
12	0	6	32	97	100	6	34	82	85	6	30	90	78	4	20	74	65
13	0	7	40	93	68	6	42	89	80	2	42	68	18	5	44	71	48
16	1	5	42	90	79	0	6	56	33	0	0	50	10	2	22	55	24
17	0	—	10	62	64	—	—	63	42	—	—	60	70	—	50	93	93
19	0	2	20	62	45	2	24	68	30	1	8	48	8	1	10	52	31
21	0	—	—	99	97	—	—	84	79	—	—	83	70	4	28	67	46
23	0	4	36	93	81	4	22	59	10	8	40	90	83	2	20	73	41
24	1	4	48	81	48	1	0	53	10	1	14	60	38	0	6	50	20
26	1	—	—	72	47	1	8	59	19	0	2	40	11	1	8	50	8
28	1	4	42	89	70	3	48	79	79	1	22	51	31	5	62	77	76
30	1	7	60	77	61	3	10	78	51	1	2	51	14	4	18	81	69
33	0	4	22	78	60	2	6	60	21	2	8	53	16	1	0	50	3
35	1	7	62	90	79	7	60	92	80	9	54	90	84	9	50	88	87
38	1	9	44	92	82	0	0	50	14	1	2	50	6	0	0	50	5
42	0	6	24	84	85	4	28	70	53	5	22	71	38	2	4	50	11
49	0	13	54	90	87	8	38	87	82	9	26	85	82	10	36	79	74
51	1	6	30	77	67	4	20	66	29	0	0	50	18	0	2	51	1
52	1	3	44	92	79	4	26	86	62	3	32	85	72	2	26	81	60
55	1	4	22	72	56	2	10	70	44	0	0	50	5	0	0	50	5
79	1	4	14	58	31	2	6	52	4	2	6	49	9	0	2	43	18
80	1	6	44	88	49	4	32	93	89	3	18	68	61	0	0	50	18
86	1	6	36	92	84	2	10	53	6	1	2	48	10	—	—	—	—
95	1	5	40	84	67	—	—	—	—	—	—	—	—	—	—	—	—
100	1	6	44	87	60	—	—	60	14	1	4	50	5	—	20	61	65
101	0	6	54	93	79	5	48	82	34	6	40	88	47	1	6	44	29
107	1	4	54	82	77	3	34	70	87	0	6	40	59	—	—	60	95
111	0	3	46	86	87	3	44	87	52	2	12	55	18	0	4	53	11
112	1	5	48	87	77	—	—	—	—	3	20	56	10	0	0	50	0
113	0	5	30	64	49	7	40	58	17	5	32	70	37	2	14	57	15
114	0	8	50	89	68	—	—	—	—	—	—	—	—	0	0	50	0
115	1	7	56	92	75	4	38	86	69	5	18	62	25	5	40	76	48
117	0	7	64	97	80	6	38	84	56	—	—	—	—	—	—	—	—
120	0	4	34	65	24	4	24	61	31	6	42	86	62	6	46	82	49
123	1	8	58	92	89	6	40	89	70	6	46	89	37	3	40	62	19
124	0	6	54	85	51	2	22	62	19	1	6	50	20	12	58	88	82
126	1	5	52	82	51	—	42	89	30	2	12	61	20	—	—	—	—
130	1	7	70	96	85	2	10	53	9	0	4	17	22	0	0	37	0
131	0	6	62	86	83	4	22	68	38	2	8	52	8	1	36	57	15
132	0	6	66	93	75	5	40	95	77	6	54	96	80	3	18	64	10

Appendix 11 (*cont.*)

Case	Group	Week														
		0				2				4				12			
		Mass	hdr	va1	va2	Mass	hdr	va1	va2	Mass	hdr	va1	va2	Mass	hdr	va1	va2
134	1	5	44	90	65	4	18	65	31	2	18	58	20	5	28	62	39
135	0	4	50	85	76	5	56	96	84	2	22	52	16	0	6	51	11
139	1	6	56	89	58	9	62	99	90	—	—	—	—	3	46	77	67
141	0	3	16	80	26	1	14	53	8	1	14	50	6	1	6	50	17
143	0	5	42	78	34	5	20	85	39	2	24	65	28	0	2	29	1
144	0	7	72	90	79	3	28	67	28	3	20	62	15	3	28	60	23
145	0	4	52	85	70	2	28	73	35	2	24	62	48	3	28	64	24
153	1	7	44	85	56	1	16	60	26	1	16	59	20	0	10	50	11
154	1	4	44	79	69	3	22	60	22	—	—	—	—	5	42	75	37
155	1	5	56	88	72	2	18	73	46	1	26	54	40	0	2	50	22
156	0	5	36	90	72	3	40	88	30	0	2	42	10	0	2	50	4

Key to groups: 1 = Real ECT, 0 = Sham ECT

Appendix 12: How large should our samples be?

Introduction

Once a decision has been made to run a study and collect data (be it a sample survey or clinical trial described in Chapters 2 and 5, respectively), we must decide how many (observations) sampling units are required to give a good chance of realizing the study objective. To collect too few might result in an inconclusive result; to collect far too many is wasteful of resources.

The procedures used to choose the correct sample size are more technical than the remainder of this book, and we so present the material here in an appendix. Two situations are covered: the first relates to calculating the sample size for a proportion and develops the material introduced in Sections 10.7 and 10.8; the second outlines how to calculate sample size for quantitative measures using as a model the ECT study, described in Section 5.7.

Proportions

When working with proportions, it is desirable to know in advance the optimal number of sampling units (usually patients) that are required to give a meaningful and trustworthy result. Should it be 10 patients or 10000?

There are two possible approaches to choosing the sample size:

1. to include as many patients as the available resources (cost or time) permit; or

2. to use an approach based on the desired level of accuracy in the results.

There are two main conflicting problems that arise with the first method. First, although it is true to say that larger samples will invariably result in greater statistical confidence, there is nevertheless a 'diminishing returns' effect. In many cases, the resources needed to collect very large samples might be better spent in extending the research in other directions. On the other hand, obtaining too few observations may result in an untrustworthy estimate. The greater the desired level of precision, the bigger the sample size should be.

Alternatively, there is a possible third approach, which is to balance the need for accuracy with the costs involved in using a sample of that size. This is the approach we advocate for practical purposes.

Confidence intervals provide the 'best guess' of where a true population proportion is likely to be found, based on the proportion in the sample. The breadth of a confidence interval indicates the precision or amount of trust that can be placed in each estimate. The narrower the confidence interval the greater is the precision, and vice versa.

The formula for calculating confidence intervals depends upon the sampling method. The approach outlined below assumes that a sample survey is performed using **simple random sampling**. Full details of sample size calculations for other sampling methods (e.g. stratified sampling, cluster sampling, etc.) are considered by Scheaffer *et al.* (1996).

As we show in Section 10.6, when the sample size is small relative to the size of the study population ('small' is usually taken to be less than 5 percent), a 95 percent confidence interval for a true population proportion is given by:

$$p \pm 1.96 \times S.E.$$

where the standard error, $S.E. = \sqrt{\dfrac{p(1-p)}{n}}$ and p is the observed sample proportion based on a sample of size n.

The amount of error in an interval is one-half the width of the interval. Therefore, the above equation can be thought of as $p \pm e$, where the error with respect to estimating a particular proportion is given by

$$e = 1.96\sqrt{\frac{p(1-p)}{n}}$$

The above formula can be rearranged to give an expression for calculating the sample size, n, necessary to achieve a given level of precision,

$$n = \frac{1.96^2 p(1-p)}{e^2}$$

The sample size depends upon the true population proportion, p; the product $p(1-p)$ being greatest when $p = 0.5$. This means that for a given error, e, sample size will be greater when $p = 0.6$ than when $p = 0.8$.

If the sample has been drawn from a population of known size N, a smaller estimate of n is obtained using the finite population correction factor, as described in Section 10.8. In this case, the sample size is calculated using

$$n = \frac{1.96^2 p(1-p)}{e^2 + 1.96^2 p(1-p)/N}$$

In practical situations, we do not know the true population proportion, p. However, an approximate sample size can be found by replacing p with an estimated value. Such an estimate might be available from a previous survey or if no such prior information is available, use $p = 0.5$ to provide a conservative estimate of sample size (one that is likely to be larger than required).

Example

The objective of this survey is to assess the effectiveness of asthma care within a single GP practice. The measure of clinic effectiveness is the proportion of patients using peak flow meters. The study population comprised all the 500 asthmatic patients registered with the clinic ($N = 500$). What sample size should be used to achieve a 95 percent confidence interval for the population proportion if the error was to be limited to 5 percent?

The likely population proportion based on a recent Medical Audit Advisory Group report was estimated as 70 percent. Using this information, i.e. $N = 500$, $e = 0.05$, and an estimated $p = 0.70$,

$$n = \frac{1.96^2 p(1-p)}{e^2 + 1.96^2 p(1-p)/N} = \frac{1.96^2(0.70)(0.30)}{0.05^2 + 1.96^2(0.70)(0.30)/500} - 196$$

Thus, if the true population proportion is around 70 percent, a sample size of 196 patients yields a 95 percent confidence of width 10 percent ($= 2e$). This sample size is almost 40 percent of the sample population! If the tolerable error (e) is increased, the sample size is reduced. For example, if the error is relaxed to 10 percent (95 percent confidence interval width $= 20$ percent), the required sample size is $n = 70$.

In the absence of prior information concerning the true population proportion, $p = 0.5$ would have been used. For a 95 percent confidence interval of breadth 10 percent the required sample size would have been 217 patients. The required sample size for a higher tolerable error of 10 percent is 80.

There are very few surveys in which a single outcome measure can realize the study objective. It is also inefficient to take only one piece of information from a sample of, say, patient records. In the asthma clinic effectiveness study, it would be sensible to collect details of additional measures, for example, hospital attendance due to asthma, lung function test performed in last 12 months, family history of asthma recorded, etc.

The question now arises as to which of these measures should be used as the basis for the sample size calculations. The easiest solution, bearing in mind that the surrogate measures will be correlated, is to use the measurement whose estimated population proportion is closest to 0.5, as the confidence interval will be widest for this proportion.

The usual reaction when statistical methods are used as the scientific basis for determining sample size is shock at the unexpectedly large number of patients usually required. The next step is to forget about the science and simply proceed with a smaller survey.

The advantage of using the above formulae is that they provide insight into both the factors that affect sample size calculation and give a realistic evaluation of what can be achieved when different numbers of patients are included in a study.

Calculating sample size for a quantitative measure

The objective of the clinical trial described in Section 5.7 is to assess the effectiveness of Electroconvulsive Therapy (ECT) in treating patients with depressive illness. The chosen trial design is a 'double-blind' group comparative study, in which patients receive either Real ECT or Sham ECT. How many patients are needed in each group to provide a reliable comparison? To answer this question, we need:

1. to decide which statistical test to perform once the data has been collected;

2. estimates, based on previous data or experience, of the

   - variance (s^2) of the difference between treated groups, and

   - the size of clinically relevant difference between groups which, if it was found, would provide unequivocal evidence of a real effect (Δ).

3. Significance level of the test (usually 0.05).

4. Statistical power of the test. This means that if we observe an actual difference of (Δ) between the two treatment then a given percentage, typically 80 percent, of the time, we would like the result of the statistical test to be statistically significant (more details of 'statistical power' are given in Section 11.9). For practical purpose a statistical power of 80 percent is usually adequate.

The basic tenet is to choose a sample size that marries both 'statistical' and 'clinical' significance.

The formula for calculating the number of patients, n, in each group in order to compare two groups is

$$n = \frac{2s^2}{\Delta^2} \times Q$$

Typical values for Q associated with both the significance level and a statistical power of 80 percent are shown in the table below. Note that the value of Q increases as the significance level of the test decreases.

Table A12.1   Values of Q to be used in the formula for calculating the required number of patients for a quantitative measure

Significance level of test	Values of Q associated with 80 percent Power
0.1	6.2
0.05	7.9
0.02	10.0
0.01	11.7

Example

In the ECT study we compare the 'real' and 'sham' ECT treated groups on the basis of Week 4 Hamilton Depression Rating Scale (HDR) percentage changes from baseline using a two sample t-test (see Example 15.5, Section 15.11). Calculate the number of patients required to compare the real and sham ECT treatment groups for a test at the 0.05 significance level having 80 percent power.

In Table 10.2, the value of Q in the 0.05 significance level row, and 80 percent power column, is 7.9. We have an estimate the estimate of variance from the

Northwick Park study of 1600 (s^2), and a difference between groups of at least 30 percent (Δ) is unequivocal. For example, if over the four-week treatment period, the HDR score in the real ECT group reduced by an average of 60 percent compared to baseline, and the average reduction from baseline in the sham ECT group was 30 percent, we would regard this 30 percent difference between treatments as overwhelming evidence of a treatment effect.

The number of patients require in each treatment group is

$$n = \frac{2s^2}{\Delta^2} \times Q = \frac{2 \times 1600}{30^2} \times 7.9 = \frac{2 \times 1600}{900} \times 7.9 = 28.1 = 28$$

Since we cannot have 0.1 of a patient, it is usual to round the estimated sample to the nearest whole number. Here 28 patients in the real and sham ECT groups will give an 80 percent chance of deeming a difference between groups of 30 percent statistically significant, with respect to HDR Week 4 percentage differences from baseline, in a test performed at the 0.05 significance level.

BIBLIOGRAPHY

Ahlbom, A. and Norell, S. (1984) *Introduction to Modern Epidemiology*. Epidemiology Resources Inc. ISBN 0–917227–00–X

Brandon, S., Cowley, P., McDonald, C., Neville, P., Palmer, R. and Wellstood-Eason (1984) Electroconvulsive therapy: results in depressive illness from the Leicestershire Trial. *British Medical Journal*, **288**, 22–25.

Bhargava, S.K., Ramji, S., Kumar, A., Mohan, M., Marwah, J. and Sachdev, H.P.S. (1985) Mid-arm and chest circumferences at birth as predictors of low birth weight and neonatal mortality in the community. *British Medical Journal*, **291**, 1617–1619.

Cimprich, B. and Ronis, D.L. (2001) Attention and symptoms distress in women with and without breast cancer. *Nursing Research*, **50**, 86–94.

Department of Health (1993) *Changing Childbirth. Part 1: Report of the Expert Maternity Group*. HMSO, London.

Department of Health (2001) *Department of Health – Medicines Control Agency (19 July 2001) Extended prescribing of prescription only medicines by independent nurse prescribers, Department of Health, London.*

English National Board of Nursing, Midwifery and Health Visiting (1999) *Strengthening Pre-registration Nursing and Midwifery Education. Sectional General Curriculum Guidance and Requirements for Pre-registration Nursing and Midwifery Programmes*. ENB, London.

Fowler, J., Cohen, L. and Jarvis, P. (1998) *Practical Statistics for Field Biology 2nd Edition*. Wiley. ISBN 0 471 98296 2.

Institute of Manpower Studies (March 1993) *Mapping Team Midwifery*.

Johnstone, E.C., Lawler, P., Stevens, M., Deakin, J.F.W., Frith, C.D., McPherson, K. and Crow, T.J. (1980) The Northwick Park Electroconvulsive Therapy Trial. *Lancet*, **ii**, 1317–1320.

Jones, B. and Kenward, M.G. (1989) *Design and Analysis of Cross-Over Designs*, Chapman & Hall, London. ISBN 0 412 30000 1.

Pocock, S.J. (1983) *Clinical Trials: A Practical Approach*. Wiley, New York. ISBN 0 471 90155 5.

Scheaffer, R.L., Mendenhall, W. and Ott, R.L. (1996) *Elementary Survey Sampling*, 5th Edition. Duxbury Press. ISBN 0 534 24342 8.

United Kingdom Central Council for Nursing, Midwifery and Health Visiting (1999) *Fitness for practice. The UKCC commission for Nursing and Midwifery Education*. UKCC, London.

Whitehead, J. (1982) *The Design and Analysis of Sequential Clinical Designs*. Ellis Horwood, Chichester.

INDEX

accuracy, 22
AGPAR scale, 16
analysis of variance, *see* ANOVA
ANOVA, 175–185
  F-test in, 180
  interaction in, 180
  Tukey test in, 181
arcsine transformation, 92
arithmetic mean, 57
association, 116, 122
average, 58
  comparison of, 158–174
axis, x and y, 35

bar graph, 30
bias, 9, 42, 53
bimodal distributions, 63
bivariate data, 27, 131
blind trials, 42

calculators, 2, 9, 66, 71
case control studies, 51
central limit theorem, 96
central tendency, 58
chi square, 115
circle graph, 35
class mark, 26
clinical trials, 39–46
  blind, 41
  design, 40
cluster sampling, 12
coefficient, 19
  of determination, 136, 151
  regression, 143
  of variation, 72
cohorts, 50
  cohort studies, 50
computers, 2, 9, 14
confidence intervals, 91

confounding, 55
contingency table, 121
continuous variables, 15, 32, 58
control groups, 39, 48, 54
  historical, 42
correlation, 131, 132
  coefficient, 132
  product moment, 134
  Spearman rank, 137
count and count data, 6, 15, 21, 26
critical probability, 76, 107
cross-over studies, 40
curved relationships, 152

data, transformation of, 91
degrees of freedom, 67, 71, 86
dependent variable, 143
derived variables, 19
descriptive statistics, 13
deviation from the mean, 66
discontinuous variables, 15
disease, measurement of, 48
distribution, frequency, 24, 32
dot plot, 29, 31, 65

error, sampling, 9, 95, 108
  standard, 97
  type 1 and type 2, 113, 175
estimates, 96–106
ethical issues, 43
expected frequency, 80, 115–129

F-test, 180
frequency, 23
  class, 23, 24
  aggregation of, 24
  curve, 34
  distribution, 24, 32
  expected, 80

frequency (*cont.*)
  polygon, 33
  proportional, 76
  table, 22–25

goodness of fit, 115, 120
gradients, 142
group comparative studies, 40
grouped data, 23,

histogram, 32
homogeneity, tests for, 115, 118
hypothesis, experimental, 107
  null, 108–109
  statistical, 108

incidence, 48
independence, 115, 117
  tests for, 115
independent variable, 143
inferential statistics, 67
interaction, in ANOVA, 184
interval scale, 17

Kruskal-Wallis test, 161

least squares, 146
location, measure of, 58
logarithms, 20
  logarithmic transformation, 91,
    139

Mann-Whitney *u*-test, 158
matched observations, 158, 164
mean, arithmetic, 57
  comparison between, 100, 168
  difference between, 100
  distribution of, 95
  of grouped data, 59
  relationship with median and mode,
    64
measurement, scales of, 15–17
median, 60
modal class, 62
mode, 62

nominal scale, 15
non-parametric statistics, 14, 91

normal curve/distribution, 81–84, 111
null hypothesis, 108–109

observations, 6
  grouped, 59
  matched, 158, 164
  precision of, 21
  tied, 18
observer bias, 9
one-tailed tests, 110
ordinal scales, 16, 62

parameter, 13
parametric statistics, 14, 113
percentage, 19, 35
pie graph, 35
placebo, 41
population, definition of, 5–6
  study, 8
  target, 8
power of a test, 114
precision, 21, 62
prevalence, 48
probability, 14, 76
  compound, 76
  critical, 76, 108
  density, 78
  distribution, 78, 79
processing data, 15–27
product moment correlation coefficient,
    134
proportions, 19, 25, 48, 92
  standard error of, 103
proportional frequency, 76
psychological effects, 41

quota sampling, 11

random numbers, 9
  samples, 9, 42
range, 25, 66
rank scale, 16
ranked observations, 18, 91, 114, 152
rates, 20
  birth, 20
  incidence, 20, 51
  mortality, 20
ratios, 19

ratio scale, 17
reciprocal transformation, 152
rectilinear equation, 142
regression analysis, 141
   coefficient, 143
   least squares, 146
   line, 141
   simple linear, 147
risk, relative, 51, 53

sample, 6
   cluster, 12
   design, 8
   quota, 11
   simple random, 9, 11, 105
   strategy, 7
   stratified, 10
   systematic, 9
sampling distribution, 95
   error, 9, 95, 108
   strategy, 7
scales of measurement, 15
scattergram, 35, 45
sequential studies, 41
simple linear regression, 147
skew/skewed distribution, 91
square root transformation, 156
standard deviation, 66–70, 83
standard error, 97
   of a difference, 101, 102
   of the mean, 97
   of a proportion 104
statistics, descriptive, 13, 45
   inferential, 13
   and parameters, 13
   sample, 13
   test, 110
stratified sampling, 10, 55
sum of squares, 68, 177

systematic sampling, 9

t, 86, 99
   -distribution, 86
   -test, 170
      for matched pairs, 171
test, statistical, 84, 107
   statistics, 110
tied observations/ranks, 18
transformation of data, 20, 91
   arcsine, 92, 156
   logarithmic, 91, 139, 152
   reciprocal, 152
   in regression, 155
   square root, 156
trials, clinical, *see* clinical trials
Tukey test, 181
two-tailed tests, 110

unit, sampling, 5
U-test, Mann-Whitney, 158

variability, 65–73
variable, 5
   derived, 19
   dependent and independent, 143
variance, 68–69
   analysis of, *see* ANOVA
variation, coefficient of, 72

Wilcoxon's test for matched pairs, 164

Yates correction, 119

z, 83, 169
   -score, 83
   -test, 168–170
Zero observations, in transformation,
   91